The Scientific Sublime

The Scientific Sublime

Popular Science Unravels the
Mysteries of the Universe

ALAN G. GROSS

Oxford University Press is a department of the University of Oxford. It furthers
the University's objective of excellence in research, scholarship, and education
by publishing worldwide. Oxford is a registered trade mark of Oxford University
Press in the UK and certain other countries.

Published in the United States of America by Oxford University Press
198 Madison Avenue, New York, NY 10016, United States of America.

Library of Congress Cataloging-in-Publication Data
Names: Gross, Alan G., author.
Title: The scientific sublime: popular science unravels the
mysteries of the universe / Alan G. Gross.
Description: New York, NY : Oxford University Press, [2018] |
Includes bibliographical references and index.
Identifiers: LCCN 2017034685 (print) | LCCN 2017039862 (ebook) |
ISBN 9780190637781 (updf) | ISBN 9780190637798 (epub) |
ISBN 9780190637804 (online course) |
ISBN 9780190637774 (cloth : alk. paper)
Subjects: LCSH: Science—Social aspects.
Classification: LCC Q175.5 (ebook) | LCC Q175.5 .G7568 2018 (print) | DDC 501—dc23
LC record available at https://lccn.loc.gov/2017034685

9 8 7 6 5 4 3 2 1

Printed by Sheridan Books, Inc., United States of America

For Bradley Benson and Aleksandr Lazaryan
Two extraordinary and caring physicians

CONTENTS

ACKNOWLEDGMENTS

I am the author of this book; my name appears alone on the title page. But if you are lucky—and I was lucky—you are a member of a team dedicated to your assistance. For their insightful efforts at improving my draft I would like to thank Randy Allen Harris and the two other reviewers who, regretfully, chose to remain anonymous. Nothing can adequately repay this gift of constructive criticism. I would also like to thank the ten authors I have chosen as my subject. One of my pre-publication reviewers was kind enough to say that my chapters read like pieces in the *New Yorker*. However true this may be, one fact must be noted: it is impossible for one's sentences not to improve when writing in the shadow of these masters and mistresses of English prose.

I would also thank my editor Hallie Stebbins for her continued support and my former editor, Peter Ohlin, who referred me to her. I would like also to pay tribute to Arthur Fine and to remember David Hull, who together taught a middle-aged man philosophy of science, and the late Joseph Williams, who taught me how to write. Last but not least, I would like to thank my long-time co-author, Joseph Harmon, who started this project with me, helped it along manfully, but decided in the end to bow out. I shall miss him; I miss him already.

It goes without saying that I alone am responsible for any errors in fact or in judgment.

The Scientific Sublime

CHAPTER 1 | Isn't Science Sublime?

The most beautiful thing we can experience is the mysterious. It is
the source of all true art and science. He to whom the emotion is a
stranger, who can no longer pause to wonder and stand wrapped in
awe, is as good as dead—his eyes are closed. The insight into the
mystery of life, coupled though it be with fear, has also given rise to
religion. To know what is impenetrable to us really exists, manifesting
itself as the highest wisdom and the most radiant beauty, which our
dull faculties can comprehend only in their most primitive forms—
this knowledge, this feeling is at the center of true religiousness.

—ALBERT EINSTEIN[1]

We were all overwhelmed with admiration and astonishment; indeed,
our surprise was so great that we had to repeat the experiment just to
be absolutely sure we were not mistaken.

—FLORIN PÉRIER, in a letter to Blaise Pascal[2]

FOR OVER A half century, popular science books have been embraced enthu-
siastically by the welcoming public, from Richard Dawkins on evolution
to Brian Greene on string theory. But while shelf upon shelf of books
of popular science exist, only one book exists on these books, Elizabeth
Leane's *Reading Popular Physics*.[3] Perhaps that's because no other book
is needed; perhaps there is no more mystery to solve, no conundrum to
unravel. Take *A Brief History of Time*: it is selling far better than *Gone
with the Wind*, apparently with good reason: it is a better read. A reviewer
on Amazon opines: "Stephen Hawking is an established scientific genius, but this book establishes him as a brilliant writer—an extremely rare,
yet valuable combination." A blog critic pronounces his verdict: "*A Brief
History of Time* is far more than a science book. It's one of the renaissance

books that is so seminal to the notion of who we are, and where we might be in the next 50 years, that it should be required reading for every person from high school on. If that seems like a big ask you've got the wrong idea about this book. It's light and easy and fun, full of subtle humor and provocative notions."[4] These are views about a book chock-full of abstruse ideas strenuously avoided in their school years by all but future physicists. The universal attraction of such books is the mystery I would like to solve, the conundrum I would like to unravel.

Jon Turney, a scholar of popular science and former editor of Penguin Books, questions whether such a book can be written: "At some point," he says, "one must ask if it is possible . . . to consider the whole ensemble of books. I have my doubts. Even books on the same topic, quantum physics say, are tremendously diverse, in style, level, approach, and in which genres they draw on."[5] Turney is not totally despairing of success; he suggests that potential authors see popular science books as symptoms of larger forces in our culture. I intend to act on Turney's suggestion. I acknowledge the diversity of style, level, and approach that Turney sees as an obstacle to a comprehensive account. But I attribute this diversity not to a difference in goals but to differing literary talents and to different takes on what science is and what it can accomplish. However different their skills and their subject matter, these writers are in the business of generating in their readers a sense of wonder at a nature whose workings science, and only science, can comprehend.

To create this transcendental effect, this sense of emotional and intellectual uplift, these writers rely on a form of experience well established in the West: the sublime. We experience the sublime in literature when we read John Milton's sonnet on his blindness; we experience it in nature when we visit the Grand Canyon; we experience it in science when we realize that one short equation—$e = mc^2$—led to Hiroshima and Fukushima. Turney agrees that "science writers evoke their most telling effects by evoking the sublime"[6]; however, he reserves this effect for the heightened prose of isolated passages. This limitation also characterizes Marjorie Hope Nicolson's pioneering classic, *Mountain Gloom and Glory: The Development of the Aesthetics of the Infinite*.[7] For her, the sublime is an aspect of Romantic poetry, of the Romantic vision of the world. I differ from Turney and Nicholson in generalizing the sublime to whole works. I contend that heightened passages in the works of popular science I have chosen are outward signs of the sublimity that inheres in their structure, a spirit that informs their every aspect.

My views are not without distinguished predecessors. In *Victorian Sensation*, James Secord explores the impact on its many readers of Robert

Chambers's bestselling *Vestiges of the Natural History of Creation*, a book described at the time as "at once scientific, metaphysical, original, daring, and sublime."[8] *Vestiges* is the grandfather of the evolutionary epic, a book Secord links to authors I have chosen to analyze: "It is through reading the successors to *Vestiges* that we make sense of our origins and potential futures. In best-sellers ranging from Stephen Hawking's *Brief History of Time* (1988) to Steven Pinker's *The Language Instinct* (1994), readers trace stories that start from swirling clouds of cosmic dust and end with the emergence of mind and human culture."[9] Writing of the evolutionary epic as a central theme of 19th-century science, Bernard Lightman cites reviewers for whom the work of Robert Ball "excite[d] astonishment" by presenting his readers with "a fresh appreciation of the sublimity of that scheme of creation in which our earth plays a small though dignified part."[10] In his *Cosmos*, Alexander von Humboldt says of the epic of evolution that

> beginning with the depths of space and the regions of the remotest nebulae, we will gradually descend through the starry zone to which our solar system belongs, to our own terrestrial spheroid, circled by air and ocean, there to direct our attention to its form, temperature, and magnetic tension, and to consider the fullness of organic life unfolding itself upon its surface beneath the vivifying influence of light.[11]

In *On Human Nature*, Edward O. Wilson identifies the epic as the narrative form of scientific materialism, a tale of "the evolution of the universe from the big bang of fifteen billion years ago through the origin of the elements and celestial bodies to the beginnings of life on earth."[12] Of such 20th-century embodiments of the evolutionary epic as Steven Weinberg's *The First Three Minutes*, Martin Eger says that they tell *"one and the same story . . .* the story of evolution *. . .* a new epic [so vast] and so detailed that no one book can encompass it."[13]

Another of my predecessors is Elizabeth Kessler for whom the images taken by the Hubble Space Telescope evoke an astronomical sublime, the vastness of the cosmos causing us to "reach out with our minds and our aesthetic and our imaginations and our sense of wonder and our sense of curiosity. . . and grab hold of the universe and make it ours."[14] Finally, we have David Nye, for whom there is a technological sublime, one manifested in objects as different as the Golden Gate Bridge and the atom bomb, and in events as different as the 1939 World's Fair and the Apollo flight to the moon. Nye's work is an application of the insight of his mentor, Leo

Marx, for whom "the awe and reverence once reserved for the Deity and later bestowed on the visible landscape is directed toward technology, or rather the technological conquest of matter."[15] To me, the sublime can be as abstruse as the Standard Theory in physics and as concrete as the Large Hadron Collider; the sublime in biology can be as abstract as evolutionary theory or as concrete as a dinosaur footprint.

The Literary Sublime

The sublime takes its first step toward cultural immortality in an essay written around the first or second century of our era and attributed, in all probability misattributed, to Longinus. *On the Sublime* observes that sublimity in poetry or prose "consists in a consummate excellence and distinction of language."[16] These works do not "persuade the audience but . . . transport them out of themselves,"[17] by drawing their attention away "from reasoning to the enthralling effect of the imagination."[18] Such works lift them "near the mighty mind of God."[19]

On the Sublime mysteriously disappears for over fourteen hundred years, only to resurface two-thirds intact, published in Basel in 1554. By the end of the 18th century, a steady stream of Greek editions and Latin translations follows. [20] Although turned into English as early as 1652, it is not until 1704, in John Dennis's *The Grounds of Criticism in Poetry* that *The Sublime* finds its way into English literary criticism. Dennis defines the sublime as "nothing else but a great thought, or great thoughts moving the soul from its ordinary situation by the enthusiasm that naturally attends them."[21] It is literature not life that creates this effect:

> For the spirits being set in a violent emotion, and the imagination being fired by that agitation; and the brain being deeply penetrated by those impressions, the very objects themselves are set as it were before us, and consequently we are sensible of the same passion that we should feel from the things themselves. For the warmer the imagination is, the less able we are to reflect, and consequently the things are the more present to us of which we draw the images; and therefore where the imagination is so inflamed as to render the soul utterly incapable of reflecting, there is no difference between the images and the things themselves.[22]

Dennis shares with his fellow Britons Joseph Addison, John Baillie, Joseph Priestley, and Hugh Blair the view that the literary sublime applies to isolated passages, not to whole literary works. Longinus is the source of this exclusion: the overwhelming number of his examples are indeed isolated passages, drawn from the works of Homer, the Greek tragedians, and Genesis. Take this episode from the *Iliad* in which the gods intervene in the battle for Troy. In Pope's translation—Pope at his iambic best—the sound echoes the sense:

> The mountain shook, the rapid streams stood still:
> Above, the sire of gods his thunder rolls,
> And peals on peals redoubled rend the poles.
> Beneath, stern Neptune shakes the solid ground;
> The forests wave, the mountains nod around;
> Through all their summits tremble Ida's woods,
> And from their sources boil her hundred floods.
> Troy's turrets totter on the rocking plain;
> And the toss'd navies beat the heaving main.
> Deep in the dismal regions of the dead,
> Th' infernal monarch rear'd his horrid head,
> Leap'd from his throne, lest Neptune's arm should lay
> His dark dominions open to the day,
> And pour in light on Pluto's drear abodes,
> Abhorr'd by men, and dreadful ev'n to gods.[23]

Longinus tells us why this passage elicits awe. It is Homer's lofty language combined with a conceptual grandeur: "The earth is split to its foundations, hell itself laid bare, the whole universe sundered and turned upside down; and meanwhile everything, heaven and hell, mortal and immoral alike, shares in the conflict and dangers of that battle."[24] Inspired by this inspired analysis, the young Edward Gibbon, not yet the great historian of ancient Rome, says that its author "tells me his own feelings upon reading it; and tells them with such energy, that he communicates them. I almost doubt which is most sublime, Homer's Battle of the Gods or Longinus's apostrophe . . . upon it."[25]

The restriction to isolated passages is not long honored. Even Hugh Blair admits that "some, indeed there are, who, by a strength and dignity in their conceptions, and a current of high ideas that runs through their whole composition, preserve the reader's mind always in a tone nearly allied to the sublime; for which reason they may, in a limited sense, merit the name of continued sublime writers."[26] And Gibbon notices that there was no reason to hesitate on Longinus's account. Longinus thinks of the whole of

Book XX of the *Iliad* as sublime: "The Battle of the Gods is worthy of everything Longinus says of it. It would be difficult to find another example which reunites so thoroughly every part of the sublime both as to thoughts and language."[27] Gibbon does not notice that Longinus himself thinks of the whole of the *Iliad* as sublime: "The consistent sublimity which never sinks into flatness, the flood of moving incidents in quick succession, the versatile rapidity and actuality, brimful of images drawn from real life."[28]

From the Literary to the Natural Sublime

It is impossible to overestimate the importance and popularity of Addison and Steele's periodical *The Spectator.* Addison's estimate of 60,000 readers is not an exaggeration.[29] Trivial today, this figure represents 10% of the population of 18th-century London, the equivalent of a YouTube video gone viral.[30] Moreover, the 18th century saw an astonishing fifteen editions or reprints of the collected essays. There were translations into French and German. Other works promoting the natural sublime were also popular: Blair and Lord Kames went into seven editions, and Edmund Burke into six. Burke was translated into French and into German. In the last decade of the century, Lord Kames was translated into German and reprinted in America. The natural sublime was on a roll.

What is the natural sublime? To Addison, "when we look on such hideous objects [as a precipice at a distance], we are not a little pleased to think we are in no danger of them. We consider them, at the same time, as dreadful and harmless; so that the more frightful appearance they make, the greater is the pleasure we receive from the sense of our own safety."[31] So long as we are safe, a storm fills us with an "agreeable horror."[32] Although Addison clearly thought of himself as breaking new ground in defining a natural sublime, Longinus had long ago anticipated this expansion:

> It is by some natural instinct that we admire, surely not small streams, clear and useful as they are, but the Nile, the Danube, the Rhine, and far above all, the sea. The little fire we kindle for ourselves keeps clear and steady, yet we do not therefore regard it with more amazement than the fires of Heaven, which are often darkened, or think it more wonderful than the craters of Etna in eruption, hurling up rocks and whole hills from their depths and sometimes shooting forth rivers of that pure Titanic fire. But on all such matters I would say only this . . . it is always the unusual which wins our wonder.[33]

Edmund Burke deepens Addison's analysis. He feels that pain and terror are the only sources of the sublime, pain and terror transformed—pain not painful and terror not terrifying—and both evoking a positive emotion akin to pleasure. Burke calls it *delight*:

> If the pain and terror are so modified as not to be actually noxious; if the pain is not carried to violence, and the terror is not conversant about the present destruction of the person, as these emotions clear the parts, whether fine or gross, of a dangerous and troublesome encumbrance, they are capable of producing delight; not pleasure, but a sort of delightful horror, a sort of tranquility tinged with terror; which, as it belongs to self-preservation, is one of the strongest of all the passions. Its object is the sublime. Its highest degree I call *astonishment*; the subordinate degrees are awe, reverence, and respect, which, by the very etymology of the words, show from what source they are derived, and how they stand distinguished from positive pleasure.[34]

To Burke, astonishment is a psychological state in which the mind is so filled with a sublime object that while the experience lasts, it can entertain no other thought or image. The eruption of Vesuvius or a stormy coast in Brittany possesses "the great power of the sublime, that, far from being produced by them, it anticipates our reasonings, and hurries us on by an irresistible force."[35] To this psychological portrait, John Baillie adds that the sublime varies from person to person,[36] and that the effect depends on novelty or surprise.[37]

Hugh Blair expands the scope of the sublime. The terrible is not its only source; also sublime are "the magnificent prospect of wide extended plains, and of the starry firmament [and] the moral dispositions and sentiments, which we view with high admiration."[38] This notion of moral sublimity, which will become especially important to Immanuel Kant, harks back to Longinus: "How grand, for instance, is the silence of Ajax in the Summoning of the Ghosts, more sublime than any speech?"[39] Longinus is referring to a scene in the *Iliad* in which Odysseus, meeting Ajax in the underworld, apologizes for conduct that led to his suicide. But "while yet I speak, the shade disdains to stay, / In silence turns, and sullen stalks away."[40] It is this form of sublimity to which Lord Kames refers. He quotes from the third part of Shakespeare's *Henry VI*:

SOMERSET. Ah! Warwick, Warwick, wert thou as we are,
We might recover all our loss again.
The Queen from France hath brought a puissant power,

Even now we heard the news. Ah! couldst thou fly!
WARWICK. Why, then I would not fly! (act 5, sc. 8)

Kames comments that "such sentiment from a man expiring of his wounds, is truly heroic and must elevate the mind to the greatest height that can be done by a single expression."[41]

In British minds, natural and moral sublimity are closely allied, the former providing a path to the latter. To Addison and Anthony Ashley Cooper, first Earl of Shaftesbury,[42] the experience of natural sublimity is valuable precisely for this reason: it leads to the appreciation of God's might and his benevolence. Thomas Reid, a luminary of the Scottish Enlightenment, agrees, alluding to the Book of Revelation:

> I beg leave to add that, when the visible horizon is terminated by very distant objects, the celestial vault seems to be enlarged in all dimensions. When I view it from a confined street or lane, it bears some proportion to the buildings that surround me; but when I view it from a large plain, terminated on all hands by hills which rise one above the other, to the distance of twenty miles from the eye, methinks I see a new heaven whose magnificence declares the greatness of its author, and puts every human edifice out of countenance; for now the lofty spires and the gorgeous palaces shrink into nothing before it, and bear no more proportion to the celestial dome, than their makers bear to its maker.[43]

On the continent, Immanuel Kant synthesizes, analyzes, and deepens the idea of the natural sublime. The extent to which he relied on the writings of his British and Scottish predecessors is unclear; whatever the extent, the discussion of the sublime in *The Critique of Judgment* unfolds as if Kant knew their work.

For Kant, the sublime is an effect of the vastness and the power of nature. It is through its vastness that we experience the first category of the sublime, the mathematical. The immensity of the cosmos overwhelms us; it exceeds the capacity of our imagination. Despite our best efforts, we fail to grasp what our reason tells us must be the case:

> A tree that we estimate by a man's height will do as a standard for [estimating the height of] a mountain. If the mountain were to be about a mile high, it can serve as the unity for the number that expresses the earth's diameter and so make that diameter intuitible. The earth's diameter can serve similarly for estimating the planetary system familiar to us, and that [in turn]

for estimating the Milky Way system. And the immense multitude of such Milky Way systems, called nebulous stars, which presumably form another such system among themselves, do not lead us to expect any boundaries here. Now when we judge such an immense whole aesthetically, the sublime lies not so much in the magnitude of the number as in the fact that, the farther we progress, the larger are the unities we reach. This is partly due to the systematic division in the structure of the world edifice; for this division always presents to us whatever is large in nature as being small in turn, though what it actually presents to us is our imagination, in all its boundlessness, and along with it nature as vanishing[ly] small, in contrast to the ideas of reason, if the imagination is to provide an exhibition adequate to them.[44]

In Kant's second category, the dynamic sublime, our imagination is overwhelmed by nature's power: "Bold, overhanging, and, as it were, threatening rocks, thunderclouds piling up in the sky, and moving about accompanied by lightning and thunderclaps, volcanoes with all their destructive power, hurricanes with all the devastation they leave, the boundless ocean heaved up, the high waterfall of a mighty river, and so on. Compared to the might of any of these, our ability to resist becomes an insignificant trifle."[45] It is a power that, if we are to experience it as sublime, must not have any actual power over us, lest we feel not pleasurable agitation but real fear.

Because the sublime first manifests itself as "violent to our imagination,"[46] our moral sense must transform it into something more manageable. By placing us in contact with the Deity, this sense transcends any inadequacy we may feel in consequence of the sublime's emotional buffeting. It is our insight into the kinship between sublimity and Deity that is a source of the pleasure we experience:

Hence sublimity is contained not in any thing of nature, but only in our mind, in so far as we can become conscious of our superiority to nature within us, and thereby also to nature outside us (as far as it influences us). Whatever arouses this feeling within us, and this includes the *might* of nature that challenges our forces, is then (although improperly) called sublime. And it is only by presupposing this idea, within us, and by referring to it, that we arrive at the idea of the sublimity of that being who arouses deep respect in us, not just by his might, as demonstrated in nature, but even more by the ability, with which we have been endowed, to judge nature without fear and to think of our vocation as being sublimely above nature.[47]

This experience must be free from any inference concerning the character of the natural objects and events we view: "When we call the sight of the starry sky *sublime*, we must not base our judgment upon any concepts of worlds inhabited by rational beings, and then [conceive of] bright dots that we see occupying the space above us as being these worlds' suns, moved in orbits prescribed for them with great purposiveness; but we must base our judgment regarding it merely on how we see it, as a vast vault encompassing everything, and merely under this presentation may we posit the sublimity that a pure aesthetic judgment attributes to this object."[48] For Kant, there is no category reserved for the scientific sublime.

Expressing the Sublime

A category of experience not yet present in early modern Europe, the sublime was unavailable to the great French essayist Michel de Montaigne when he crossed the Alps in October 1580. The description betrays no experience of the sublime:

> Although most of the mountains close to us here are wild rocks, some massive, others split and broken up by the flow of the torrents, and others scaly and sending down pieces of astounding size (I believe it must be dangerous here in very stormy weather) just as elsewhere, we have also seen whole forests of pines torn up from their footing and carrying down in their fall little mountains of earth clinging to their roots.[49]

Crossing the Alps by the same Mont Cenis pass just over a century later, the poet, playwright, and literary critic John Dennis has a different reaction, the first hint that the category of the natural sublime was in the British mental repertoire, a mix of horror and pleasure:

> The ascent was the more easy, because it wound about the mountain. But as soon as we had conquered one half of it, the unusual height in which we found ourselves, the impending rock that hung over us, the dreadful depth of the precipice, and the torrent that roared at the bottom, gave us such a view as was altogether new and amazing. On the other side of that torrent, was a mountain that equaled ours, about the distance of thirty yards from us. Its craggy cliffs, which we half discerned, through the misty gloom of the clouds that surrounded them, sometimes gave us a horrid prospect. And sometimes its face appeared smooth and beautiful as the most even and

fruitful valleys. So different from themselves were the different parts of it: In the very same place nature was seen severe and wanton [that is, luxurious]. In the mean time we walked upon the very brink, in a literal sense, of destruction; one stumble, and both life and carcass had been at once destroyed. The sense of all this produced different emotions in me, viz. a delightful horror, a terrible joy, and at the same time, that I was infinitely pleased I trembled.[50]

Recollecting an Alpine trip of 1701–1703, Addison chimes in. He speaks of "a near prospect of the Alps, which are broken into so many steeps and precipices, that they fill the mind with an agreeable kind of horror."[51] In the same vein, writing toward the end of the century, Swiss scientist Horace-Bénédict de Saussure takes time from his work as a geologist to enjoy the Alpine view: "This situation savored of the strange and the terrible. I seemed to be alone on a rock, in the middle of a roiling sea, a great distance from a continent bordered by a long reef of inaccessible rock. Little by little, this cloud rose, enveloping me first in its obscurity; then, mounting above my head, I discovered all of a sudden a superb view of the lake and its cheerful banks, cultivated, covered with small towns and lovely villages.[52]

It is not only the Alps that evoked sublimity; there was also Mount Vesuvius and the Rhine Falls at Schaffhausen. The former was visited by Joseph Spence in 1732, the latter by Helen Maria Williams in 1798. Spence professes himself astounded by the contrast between the beauty of the Bay of Naples and the potential of this powerful and restless volcano: "On the right hand appeared part of the delicious bay of Naples; 'twas but turning the head, and we had a full view of all the city and the bay. The mixture of the most beautiful thing with the most horrid [the smoking crater of Vesuvius] that can well be imagined, had a very particular effect, and struck one in a manner that I never felt on any other occasion."[53]

Naturally prone to enthusiasm, Williams speaks of the Alps, which she had not visited as "those regions of mysterious sublimity, the solitudes of nature, where her eternal laws seem at all seasons to forbid more than the temporary visits of man, and where, sometimes, the dangerous passes to her frozen summits are inflexibly barred against mortal footsteps."[54] When she visits the Schaffhausen Falls, however, she is beside herself:

That stupendous cataract, rushing with wild impetuosity over those broken, unequal rocks, which, lifting up their sharp points amidst its sea of foam, disturb its headlong course, multiply its falls, and make the afflicted waters roar—that cadence of tumultuous sound, which had never till now struck upon my ear—those long feathery surges, giving the element a new aspect—that spray rising into clouds of vapor, and reflecting the prismatic colors, while it dispenses itself over the hills—never, never can I forget the sensations of that moment! when with a sort of annihilation of self, with every part impression erased from my memory, I felt as if my heart were bursting with emotions too strong to be sustained.[55]

Some visitors went further: Kant was not the only one for whom the natural sublime led to God. In 1739, Thomas Gray found himself and his companion, Horace Walpole, in the same Mont Cenis pass as Montaigne and Dennis. Walpole was repulsed by what he saw. He writes to his friend Richard West: "We were eight days in coming hither from Lyons; the four last in crossing the Alps. Such uncouth rocks, and such uncomely inhabitants! My dear West, I hope I shall never see them again!"[56] In a letter home, Gray records his very different experience:

I own I have not, as yet, anywhere met with those grand and simple works of art, that are to amaze one, and whose sight one is to be the better for: But those of nature have astonished me beyond expression. In our little journey up to the Grande Chartreuse, I do not remember to have gone ten paces without an exclamation, that there was no restraining: Not a precipice, not a torrent, not a cliff, but is pregnant with religion and poetry. There are certain scenes that would awe an atheist into belief, without the help of other argument.[57]

Helen Maria Williams, writing at the end of the century, was similarly transported in a passage that echoes Psalm 123: "As the mountains are round about Jerusalem, so the Lord is round about his people from henceforth even forever":

Lofty mountains are in scripture called the mountains of God; and although worship in high places has been stigmatized as idolatry, yet surely, if the temple, which best delights the Supreme Being, be a temple not made with hands, that which next to the pure and innocent heart is most worthy of his sublimity are the summits of those everlasting mountains, the faint but nearest resemblance on earth of his unchangeableness and eternity.[58]

Poetry was another medium in which the sublime experience of nature was expressed. We may safely pass over the mediocrity of George Keate's "The

Alps" and Helen Williams's "Hymn Written among the Alps."[59] But not all verse on the natural sublime falls short of the literary sublime. While my English cannot do justice to Albrecht von Haller's German, I can capture a whiff of his Alpine vision: "As the first rays of the sun gilded the peaks of the Alps and the mist vanished, there emerged from a thinning cloud as if on a stage the broad prospect of many nations seen all at once in a single glance. A gentle vertigo possessed me, overwhelmed by so wide a vista so suddenly revealed. It was more than my eyes could take in."[60]

Earlier in the century, on a second visit to the Alps, Thomas Gray tries to capture in Latin verse the experience he conveyed two years earlier in a letter to his friend Richard West. In translation, his ode reads in part:

O Thou, Holy Spirit of this stern place, what name soever pleases Thee (for surely it is no insignificant divinity who holds sway over untamed streams and ancient forests; and surely, too, we behold God nearer to us, a living presence, amid pathless steeps, wild mountain ridges and precipitous cliffs, and among roaring torrents and the nocturnal gloom of sacred groves than if He were confined under beams hewn from the citron tree and gleaming with gold wrought by the hand of Phidias [the Greek sculptor])—hail to Thee![61]

But it is only with the Romantics—William Wordsworth, Samuel Taylor Coleridge, Lord Byron, and Percy Bysshe Shelley—that such evocations of the natural sublime rise to the level of genius. Coleridge's "Hymn Before Sun-Rise, in the Vale of Chamouni" is based on a German poem by the Danish poet Friederike Brun, a debt that made him uncomfortable: "I involuntarily poured forth a hymn in the manner of the *Psalms*, though afterwards I thought the ideas etc. disproportionate to our humble mountains—and accidentally lighting on a short note in some Swiss Poems, concerning the vale of Chamouni and its mountain, I transferred myself thither, in the spirit and adapted my former feelings to these grander external objects."[62]

Coleridge had no more reason to hide the truth of indebtedness than Shakespeare needed to apologize for transmuting *The True Chronicle History of King Leir* into *King Lear.* Addressing Mont Blanc, Coleridge transforms German lead into English gold, the sound perfectly echoing the sense:

Who made you glorious as the Gates of Heaven
Beneath the keen full moon? Who bade the sun
Clothe you with rainbows? Who, with living flowers

Of loveliest blue, spread garlands at your feet?
God! let the torrents, like a shout of nations,
Answer! and let the ice-plains echo, God!
God! sing ye meadow-streams with gladsome voice!
Ye pine-groves, with your soft and soul-like sound!
And they too have a voice, yon piles of snow,
And in their perilous fall shall thunder, God![63]

In submitting to the sweep of Coleridge's sublime vision of sublimity, we might overlook a significant fact: he never actually visited Mont Blanc. A category of experience, it seems, now generates experiences more vivid than experience itself. [64]

Painting also partakes of the sublime. Joseph Turner revisited Schaffhausen Falls in 1802 and produced this masterpiece of light and color, reproduced in figure 1.1.

Turner specialist Marjorie Munsterberg says of the artist that he

emphasized the sublimity of the falls by his method of painting. . . . The water especially is painted with great sweeps of a paint-laden brush, which twists and turns as it describes the leaping, crashing water. The two materials become in some sense equivalent; the driving force of the paint represents, but also comes to be, the driving force of the water. The grand scale of the painting makes the roughness of the surface and the prominence of the white, foamy strokes overwhelming.[65]

FIGURE I.I Turner at the Rhine Falls.

Not only did poets and artists experience and express the natural sublime; this new category of experience was also realized in the minds of the many privileged English men and women on the Grand Tour of Europe. And grand the tour was: the spread of the sublime was far from democratic. When the average annual wage of a common laborer was £25 a year, his tour cost Francis, tenth Earl of Huntingdon, £5,700 a year.[66] While time has not eliminated such income disparities, for us the sublime has been democratized. The Grand Tour of the privileged is now the tourism those of far more modest means undertake; the language of Dennis, Burke, Kant, and Gray is now the language of brochures. Of the Grand Canyon a visitor notes "how thrilling [it] is, how big and vast it must have been absolutely when man first saw it." Others tell us of its "sheer vastness." It is "awe-inspiring." There is "nothing in England like it." We are told that a flight through the Alps

> is quite exciting. You start with the helicopter climbing north out of Zermatt, up the grassy slopes of the valley and within a few minutes you are flying just feet above the most remarkable glaciers with huge crevasses. Up further still you cross great rocky, knife-edged ridges and get wonderful views of peaks such as the Weisshorn. You then turn back towards the highlight of the trip, which is a very close fly past and round the Matterhorn. It really is amazing to be so close to such an iconic peak and to see the footprints of climbers reaching the summit. Then you fly across the glaciers and snows of the nearby Breithorn ridge in the direction of the Monte Rosa. The views really are almost overwhelming. In every direction are wonderful glaciers, deep rocky valleys, mountain peaks, and climbers and skiers.[67]

Of the last eruption of Vesuvius in 1944, Dr. Leander K. Powers says: "On Sunday night, the roars became more frequent and grumbled like a lion's roar. Streams of fire were shooting thousands of feet in the air, and the countryside was lit up for miles around. Of times the entire top of the mountain looked as if it were a blazing inferno. It's really uncanny, yet amazing to look at this phenomenon."[68] Contrast with this inflated prose the younger Pliny's account of the 79 AD eruption, so very matter of fact, so *scientific*, and all the more gripping in its objectivity:

> A cloud was rising from what mountain was a matter of uncertainty to those who looked at it from a distance: afterwards it was known to be Vesuvius whose appearance and form would be represented by a pine better than any other tree, for, after towering upwards to a great height with an extremely lofty stem, so to speak, it spread out into a number of branches; because,

as I imagine, having been lifted up by a recent breeze, and having lost the support of this as it grew feebler, or merely in consequence of yielding to its own weight, it was passing away laterally. It was at one time white, at another dingy and spotted, according as it carried earth or ashes.[69]

There is now a technical term for these largest and most violent of volcanic eruptions: scientists call them Plinian.

From the Natural to the Scientific Sublime

It is not to Kant but to the British and Scottish schools that we owe the extension of the sublime from nature and art to science. While there are some hints about this link in Baillie, Burke, and Priestley, it is Adam Smith who provides the fullest analysis. Baillie and Priestley see the scientific sublime in the achievement of universal principles. Priestley puts it well: "The sublime of science consists in general and comprehensive theorems, which, by means of very great and extensive consequences, present the idea of vastness to the mind . . . and the sciences of natural philosophy and astronomy, exhibit the noblest fields of the sublime that the mind of man was ever introduced to."[70] Burke adds that it is not only vastness that evokes the scientific sublime:

> When we attend to the infinite divisibility of matter, when we pursue animal life into these excessively small, and yet organized beings, that escape the nicest inquisition of the sense; when we push our discoveries yet downward, and consider those creatures so many degrees yet smaller, and the still diminishing scale of existence, in tracing which the imagination is lost as well as the sense; we become amazed and confounded at the wonders of minuteness; nor can we distinguish in its effect this extreme of littleness from the vast itself. For division must be infinite as well as addition; because the idea of a perfect unity can no more be arrived at, than that of a complete whole, to which nothing may be added.[71]

But it is Adam Smith in his *History of Astronomy* who provides us with the first satisfactory analysis of the psychology of the scientific sublime. To Smith, a path to the sublime opens when a spectacular natural event—an eclipse of the sun, for example—captures our attention and causes

> the imagination and memory [to] exert themselves to no purpose, and in vain look around all their classes of ideas in order to find one under which

it may be arranged. They fluctuate to no purpose from thought to thought, and we remain still uncertain and undetermined where to place it, or what to think of it. It is this fluctuation and vain recollection, together with the emotion or movement of the spirits that they excite, which constitute the sentiment properly called *Wonder*, and which occasions that staring, and sometimes that rolling of the eyes, that suspension of the breath, and that swelling of the heart, which we may all observe, both in ourselves and others, when wondering at some new object, and which are the natural symptoms of uncertain and undetermined thought. What sort of a thing can that be? What is that like? are the questions which, upon such an occasion, we are all naturally disposed to ask.[72]

Two other senses of wonder are evoked by the answer to these questions. The discovery of a pattern is the first, a sense not noticed by Smith. Its explanation evokes another sense of wonder, a sense noticed by Smith, the satisfying astonishment that greeted Ptolemaic astronomy on its introduction in ancient Greece, a system that made mathematical sense of the heavens:

If it [the system of concentric spheres] gained the belief of mankind by its plausibility, it attracted their wonder and admiration; sentiments that still more confirmed their belief, by the novelty and beauty of that view of nature which it presented to the imagination. Before this system was taught in the world, the earth was regarded as what it appears to the eye, a vast, rough, and irregular plain, the basis and foundation of the universe, surrounded on all sides by the ocean, and whose roots extended themselves through the whole of that infinite depth which is below it.[73]

Achievements such as this cannot be appreciated by ordinary folk; the problem would never have occurred to them.[74] Moreover, Smith says that they would regard the efforts to find a solution as worthless because it lacked practical application:

It is in the abstruser sciences, particularly in the higher parts of mathematics, that the greatest and most admired exertions of human reason have been displayed. But the utility of those sciences, either to the individual or to the public, is not very obvious, and to prove it, requires a discussion which is not always very easily comprehended. It was not, therefore, their utility which first recommended them to the public admiration. This quality was but little insisted upon, till it became necessary to make some reply to the reproaches of those, who, having themselves no taste for such sublime discoveries, endeavor to depreciate them as useless.[75]

It is in Newton's *Principia* that these abstruser sciences culminate:

> Can we wonder then, that it should have gained the general and complete approbation of mankind, and that it should now be considered, not as an attempt to connect in the imagination the phenomena of the Heavens, but as the greatest discovery that ever was made by man, the discovery of an immense chain of the most important and sublime truths, all closely connected together, by one capital fact [gravity], of the reality of which we have daily experience.[76]

C. P. Snow gives us a more contemporary example—the second law of thermodynamics, the principle that all closed systems eventually run down: "It has its own somber beauty; like all major scientific laws, it evokes reverence."[77]

It is not only the landscape that inspires the sublime in art; science does as well. Unlike Turner, Joseph Wright hides his brushstrokes, creating his sense of sublimity by the manipulation of light and shade in the manner of Georges de la Tour. The children in figure 1.2 experience the

FIGURE 1.2 Fascination with an orrery.

FIGURE 1.3 Metamorphosis of Messier 8. Hubble Space Telescope.

sublime especially, transfixed as they are by this orrery, the mechanical device that imitates the paths of the planets and their moons. Twenty-first-century astronomy out-Turners Turner in the image of the nebula Messier 8 depicted in figure 1.3.

It is the products of these abstruser sciences that the popularizers with whom I deal bring before their publics; it is they who solve for us the mysteries of theoretical physics, cosmology, and evolutionary biology. By doing so, they hope to generate all three senses of wonder: first, the amazement we feel when a particular phenomenon first captures our attention; second, our astonishment at a pattern in nature only science can perceive, one that includes the phenomenon in question; third, the scientific explanation of this pattern.

This Book

This book concerns a little explored cultural phenomenon: the wave of books and essays on popular science that begins in the latter half of the 20th century. I have chosen ten representative scientists, all writers of best-sellers, and all, except for Steven Weinberg, celebrities: Richard Feynman, Stephen Hawking, Brian Greene, Lisa Randall, Richard Dawkins, Steven Pinker, Stephen Jay Gould, Rachel Carson, and E. O. Wilson. Only Gould is given two chapters. Unlike the others, he has two large and largely distinguished bodies of work in two genres: books and essays. In each case, I emphasize the author's ability to create works that convey complex science clearly to the general public. My analysis explores the ways in which this feat is accomplished, in which the structure of arguments and the choice of words and images convey physics and biology across the gulf that separates professional from general understanding. While their literary artistry accounts for the comprehensibility of these books and essays, it does not account for the intrinsic appeal of their subject matter, most of which has no conceivable application in everyday lives. These works are the opposite of self-help books. They appeal not to our practical side but to the fact, that, as philosopher Martin Heidegger reminds us, we are thrown into the world, willy-nilly, and forced to find our way among mysteries we cannot hope to understand on the basis of experience. It is no accident that these books deal with questions we cannot fail to ask: How did the universe begin? Where did human beings come from? What is the proper relationship between us and other living things? What is the origin of language? This set of questions applies even to those few works that insist on their everyday relevance: Gould's *Mismeasure of Man* and Carson's *Silent Spring*.

The authors I have chosen share a common goal, the creation of a vast communal epic, a set of variations on the theme of origins: the origin of matter, of the universe, of living things, of human beings and the world they inhabit, an account of genesis, an account that competes with Genesis. Their task is to reveal hidden patterns in the natural world and to infer the laws that create them. Their ventures are the sources of the scientific sublime described by Adam Smith and Edmund Burke, the results of journeys into the realms of the impossibly vast and the incredibly small, the cosmos and the quark, the biosphere and the gene. These theories generate our astonishment that so much can be explained by so little, that vast and pervasive operations in the natural world can be reduced to a formula: force equals mass times acceleration in physics the Lotka and Volterra predator-prey equations in biology. For scientists, the experience of the sublime is the discovery of these patterns and laws; for the readers of their books and essays, it is vicarious experience of these discoveries that entices and enthralls, the product of literary talent.

This book falls into two parts. In Part 1, I deal with the physicists Richard Feynman, Steven Weinberg, Brian Greene, Lisa Randall, and Stephen Hawking. The chapters are ordered from well-founded to more speculative physics. Feynman's is a consensual scientific sublime, based on what we certainly know. By contrast, Weinberg traffics in a conjectural sublime, a reasonable extrapolation of what we know. His first popular science book is *The First Three Minutes*. Why the first three minutes? Because during that period the universe was a dense homogeneous soup, to which Weinberg can apply the existing laws of physics. Brian Greene and Lisa Randall are united in their speculative bent; they spend considerable effort in describing progress toward a theory of everything, that is, everything material. Randall's focus is theories involving dimensions of space beyond the three of everyday experience. She also brings to bear her considerable literary skills to describe the Large Hadron Collider, the most complicated experimental apparatus ever created, a device that she hopes will one day show that her theories are no longer just theories. Greene is even more speculative; his candidate for a theory of everything is "string theory," a mathematical formulation that replaces the dimensionless points of modern particle physics with vibrating strings so small they are forever undetectable. My last physicist, Stephen Hawking, wrote the bestselling popular science book of all time, *A Brief History of Time*; I single him out because, alone among contemporary scientists, he has the charisma of all four Beatles rolled into one. Confined to a wheelchair, he towers above his fellows; he is an embodiment of the scientific sublime, the Earl of Warwick redux.

In Part 2 I deal with five scientists deeply committed to evolutionary theory: Rachel Carson, Stephen Jay Gould, Steven Pinker, Richard Dawkins, and E. O. Wilson. The chapters are ordered according to the intensity and range of their authors' confrontations with science and society. Carson's warrior's stance is limited to her last book and concerns only the overuse of pesticides. Gould is more confrontational. His views on punctuated equilibrium take on evolutionary theory as it is generally understood; he also challenges society for its persistent misuse of intelligence tests, an abuse of science that he feels furthers racial prejudice. Pinker seems always ready to do battle. He wrestles with philosopher Jerry Fodor on the way the mind works and with psychological linguist David Rumelhart on what makes language tick. He is even brave enough, or foolish enough, to have a go at radical feminism. Dawkins is even more contentious. His theory of the selfish genes takes on the biological establishment; it also takes on all of us who are deluded enough to think that we are in charge of ourselves, that our genes, though they constrain, do not determine who we are. Wilson confronts all humanity. His sociobiology seems to deprive us of freedom of will. Apparently not, however, since he strives to impress upon us that our survival depends on preserving species diversity; he urges us to combat the dire consequences of human depredation. To Gould, Dawkins, and Wilson, the biosphere must be explained; to Carson and Wilson, it must also be protected, a deeply felt ethical commitment their science generates.

I am humbled by the literary skills of these scientist-authors; for these I have nothing but praise. Except for Rachel Carson, they are scientists of distinction, in the case of Feynman and Weinberg, scientists of great distinction. Writing about their sciences, none can be faulted, only understood and admired. Usually, when they present professional positions that are heavily contested, they say so; usually, they are scrupulous when dealing with opposing positions. But not always. When they do deviate from appropriate even-handedness, as in the case of Steven Pinker in his disputes over the origin of language and of concepts, I feel obliged to point this out. Moreover, these authors do not write only as scientists. Steven Weinberg and E. O. Wilson also write as philosophers; Steven Pinker, as a philosopher, historian, and art critic; Stephen Jay Gould writes as a historian and philosopher; Richard Dawkins, as a philosopher and a theologian. When these scientists stray far from their fields of expertise, they may criticize; they may even show contempt for what they do not trouble to understand. In these cases, I feel duty-bound to point out that their use of the scientific sublime adds a gloss to arguments they would dismiss out of hand were they measured against the standards they themselves invariably heed in their fields of expertise.

This is especially true of their effort to substitute science for God. In my final chapter, I take this antagonism up, focusing on the scientist who makes the most aggressive case, Richard Dawkins. I show that he cavalierly casts aside the achievements of two millennia of Western religious thought, not realizing that it is impossible to understand ourselves and our world without taking this body of work profoundly into consideration. This indebtedness includes the origin and evolution of science, a history rooted in Christianity and Islam. While Dawkins is the author who is most intense in his antagonism, others are in essential agreement. Whatever the place of God in the ten popular scientists I discuss, their books and essays are designed to convert their readers to a particular vision of the origin of the universe, of us, and of the cultural and social world that surrounds us, a vision in which science is the only generating and guiding force. But do these origin stories—as well-reasoned and as awe-inspiring as they are, as sublime as they are—render religion superfluous? They do so only on the condition that science can convincingly address two questions none of us can avoid: What is the meaning of life? What constitutes the good life?

PART I | The Physicists

CHAPTER 2 | Richard Feynman
| *The Consensual Sublime*

Speaking is easy. But writing is not easy.

—RICHARD FEYNMAN

RICHARD FEYNMAN WAS a fox, not a hedgehog: he did not know one big thing; instead, he knew many things. He was an inspired tinkerer, a Thomas Edison of theoretical science. Still, like Leo Tolstoy, he yearned to be a hedgehog. Feynman's vision was like Tolstoy's: "scrupulously empirical, rational, tough-minded and realistic. But its emotional cause is a passionate desire for a monistic vision of life on the part of the fox bitterly intent on seeing in the manner of the hedgehog."[1] This difference extends to method and attitude. While the great physicist Hans Bethe, Feynman's frequent working companion at Los Alamos, proceeded deliberately in any argument between them, Feynman "was as likely to begin in the middle or at the end, and jump back and forth until he had convinced himself he was right (or wrong)." It was a contest between "the Battleship and the Mosquito Boat," a small, lightly armed torpedo vessel.[2]

From 1948 to 1958, Feynman enjoyed triumph after triumph. To a former student, Koichi Mano, Feynman wrote: "You met me at the peak of my career when I seemed to you to be concerned with problems close to the gods."[3] Working on these problems, Feynman reflects a general conviction typical of successful scientists. Another scientist says what Richard Feynman might have: "There's nothing I'd rather do. In fact my boy says I am paid for playing. He's right. In other words if I had an income I'd do just what I'm doing now. I'm one of the people who has found what he wanted to do. At night when you can't sleep you think about your problems. You work on holidays and Sundays. It's fun. Research is fun. By and large it's a very pleasant existence."[4]

Problems close to the gods are their gift, but the gods are capricious. This is why for many geniuses, being a genius is a career as brief as an athlete's. For most, as for Feynman, a dreaded day arrives: the great insights stop coming. The marvelous decade having passed, Feynman tells his student Mano that he turned to "innumerable problems you would call humble." With elite scientists, the problem is general. Asked by sociologist Joseph Hermanowicz: "Do you ever doubt yourself?" an elite scientist answered: "Yes. Daily. I don't have any doubts about my technical abilities. I have doubts whether I will have another good idea."[5] It is this phenomenon that accounts for the surprising revelation that elite scientists are the most dissatisfied with their lifetime of achievement:

> Among the most striking developments among elites at the end of their careers is their change in work attitude, which, while moderately positive, becomes ambivalent. At the end, elites develop a heightened concern about their professional standing as they realize the improbability of change in the standing they have achieved. In a reversal, their overall satisfaction drops, best described as "medium-low," that reflects a disappointment in not having achieved more. In face-to-face interviews, elites solemnly communicate this void as a kind of dull, prolonged ache—as if grieving a loss— and at other times an angry bitterness. Among the scientists at the end of their careers, it was most difficult to talk with elites, arguably the highest achieving, about achievement. In another twist, elites—the most rewarded scientists—perceive their careers as not having gone as expected.[6]

The fall can be precipitous because the heights are so high. At his best, Feynman is a wonder to behold, a man who revels in his genius.

During the January 1949 meeting of the American Physical Society, a paper is read that claims to have eliminated troublesome infinities in a class of subatomic particles. A member of the audience, the great Robert Oppenheimer, rises to dismiss the claim. Is the physicist aware of a result disproving it? Does he know about Case's theory? Since this result by one of Oppenheimer's postdoctoral students is unpublished, the speaker, Murray Slotnick, is rendered speechless by this brusque dismissal. After the meeting, Feynman, who has not attended, becomes aware of the challenge Oppenheimer had posed. He consults with Slotnick; overnight, he works feverishly on the problem. As a consequence, he proves not only that Slotnick is correct but that the special case on which Slotnick has focused can be generalized: it is true across the board. Attending the presentation of Case's paper the next day, Feynman cannot resist the opportunity to dazzle and dismay. "What about Slotnick's calculation?" he asks. Slotnick is stunned: what has taken him two years Feynman has exceeded over-night: "That was the moment when I got my Nobel Prize when Slotnick told me he had been working two years. When I got the real prize it was really nothing, because I already knew I was a success. That was an excit-ing moment."[7]

This brilliant man had a brilliant second career. After the glory days of making significant contributions to theoretical physics, he labored suc-cessfully to enlighten general readers concerning science in general and quantum theory in particular. In *The Character of Physical Law*, his target is the extent to which the laws of science can be woven into a single fab-ric. In *QED: The Strange Theory of Light and Matter*, he introduces us to the bizarre micro-world that underlies the familiar world of experience. Neither of these books is meant for those who wish "to have the illusion of understanding and to catch a few buzz words to throw around at cocktail parties."[8] In the introduction to *QED*, Feynman introduces his philosophy of popular science: "People are always asking for the latest developments in the unification of this theory with that theory, and they don't give us a chance to tell them anything about what we know pretty well. They always want to know the things we don't know. So, rather than confound you with a lot of half-cooked, partially analyzed theories, I would like to tell you about a subject that has been very thoroughly analyzed."[9]

In this conviction, Feynman differs from Steven Weinberg, who is happy with sound conjecture; he differs also from Stephen Hawking, Brian Greene, and Lisa Randall, who traffic in informed speculation. Feynman's popular work is grounded in facts and theories that have passed the test of experimental verification, facts and theories not about to change tomor-row, or the decade later. Their scientific sublime is founded in plausibil-ity; his is built upon a rock, a consensual sublime. His only concession to general audiences is the almost complete absence of mathematics; that deepest of sublimes is reserved for physicists alone.

Feynman's pedagogical enterprise is not limited to *Character* and *QED*. He also shared us his insight into the cause of the *Challenger* disaster. He taught us that the consensual sublime has practical traction, that "Nature cannot be fooled."[10] In addition, his achievement as a popularizer would be shortchanged if we dismissed as mere entertainment his anecdotal rem-iniscences, *Surely You're Joking, Mr. Feynman!* and *What Do You Care What Other People Think?* Feynman's well-told stories are entertaining, to be sure, but they are also windows into his practice of science. They give us a sense of what it is like to be a fox chasing theoretical rabbits.

Feynman drilled nature down to the quantum level; he also fashioned expository tools that enabled others to learn what it is like to be one kind of physicist. What has been said about his physics lectures applies generally:

> Feynman shows on every page the sheer delight of thinking about physics, which includes reveling in its concrete, experimental detail, not just models or theories alone. He passionately reiterates the absolute centrality of experiment in physics, theorist though he was. His own high-spirited, intense voice (serious but often close to joking) brings together the worlds of things and of ideas in ever-new and amazing combinations. Reading him, we feel that physics is an engaging, compelling human experience, intimately tied up with all that is important in life, which we, too, should embrace and enjoy. Is it true, as Feynman concludes, that the physicist's way of looking at the world is "the greatest adventure that the human mind has ever begun"?[11]

The Character of Light

We begin with *QED*, Feynman's masterpiece of verbal and visual exposition. The abbreviation stands for "quantum electrodynamics," a theory that explains the "strange" behavior of tiny particles, of photons and electrons. From this interaction, many of the other laws of physics arise—those having to do with "heat, magnetism, electricity, light, X-rays, ultraviolet rays, indices of refraction, coefficients of reflection and other properties of various substances."[12] Quantum electrodynamics represents a major step forward on the path toward a theory of everything.

To explain the mysteries of the micro-world, Feynman starts with something we all have experienced—partial reflection. In otherwise transparent glass, we see a faint image of ourselves. It is through this everyday route that Feynman arrives at a central dilemma. While most photons pass through the glass, a few bounce back—the faint image of ourselves. But *which* pass through the glass? *Which* bounce back? We cannot tell; we will never be able to tell. In the micro-world, certainty is banished. Only probability reigns. Feynman makes the general point:

> I am not going to explain how the photons actually "decide" whether to bounce back or go through; that is not known. (Probably the question has

no meaning.) I will only show you how to calculate the correct *probability* that light will be reflected from glass of a given thickness, because that's the only thing physicists know how to do! What we do to get the answer to *this* problem is analogous to the things we have to do to get the answer *to every other* problem explained by quantum electrodynamics. You will have to brace yourself for this—not because it is difficult to understand, but because it is absolutely ridiculous: All we do is draw little arrows on a piece of paper—that's all![13]

These words create an effect all popular lecturers hope for: each audience member feels personally addressed by a scientist who is treating them with the respect they feel they deserve. You can almost hear the twang of Feynman's New York English. His personality shines through—brash, knowledgeable, friendly, fun-loving. Yes, he says, physicists are highly trained; yes, they have talents you do not possess; yes, they speak only in the language of mathematics. But the central concepts of quantum elec-trodynamics are simple: any intelligent person can grasp them with some mental exertion. He seems to say: We physicists have fancy names for things; we call these "little arrows" probability amplitudes. But when I translate our amplitudes into little arrows, nothing conceptual is lost, no necessary rigor is swept under the rug. Indeed, the conceptual structure of quantum electrodynamics will be the clearer for this translation, even to physicists.

Feynman's choice of words enhances his credibility with a general audience. "Absolutely ridiculous" anticipates an audience's disbelief in the absurdity of nature; "all we do" anticipates their disbelief in the rel-ative simplicity of the solution he will offer. Other indicators convey the feeling of intimate conversation. Quotation marks convey the irony that exposes the limits of quantum electrodynamics: it can tell us how; it can-not tell us why. Italics and exclamation points isolate the keywords around which the meaning of the passage crystallizes. They also convey the force of Feynman's personality. While marks on a page may be a pale second best to the experience of the real Feynman, they nevertheless convey the sense of a man who has set his genius aside to talk to us, one on one, a man whose genius as a scientist adds a gloss to his genius as a communicator.

None of this folksiness masks Feynman's serious intent. In his pref-ace to the anniversary edition of *QED*, physicist and author Anthony Zee tells us that he could make no sense of a sentence in Stephen Hawking's *Brief History of Time*. To write in Hawking's manner can, Zee implies, give readers merely the "illusion of understanding,"[14] a sin Feynman never commits. While rereading difficult passages in Hawking offers little help—the opacity seems permanently embedded—rereading Feynman is a must: understanding *QED* is work. If you pay it the compliment of serious attention, *QED* will teach you what physicists do when they explore the micro-world; Feynman will share with you an accurate analysis of the truly weird behavior of the photons and electrons that underlie and account for the world of ordinary experience. In a letter to a film executive, Feynman makes clear his attitude toward the pedagogical value of popular science:

> I am a successful lecturer in physics for popular audiences. The real entertainment gimmick is the excitement, drama, and mystery of the *subject matter*. People love to learn something, they are "entertained" enormously by being allowed to understand a little bit of something they never understood before. One must have faith in the subject and in people's interest in it.[15]

A book of only 152 pages *QED* contains a robust 93 diagrams, indicating their central role in clarifying the readers' understanding of complex physics. To demonstrate how photons, massless particles of light, create partial reflection, Feynman conducts a thought experiment. Such an experiment takes place not in the laboratory but in Feynman's imagination, an imagination that is, however, fully constrained by the facts of physics.

The fictional setup is simplicity itself: a source of light sends its beams toward a plate of glass that varies in thickness over the course of the experiment. In figure 2.1, two detectors—A at the front, B behind the glass plate—count the photons from the beam. The amount reflected from either the front or back of the glass to detector A depends on the thickness of the glass. It varies from 0% to 16%. Consequently, the probability of photons

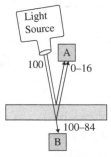

FIGURE 2.1 Experimental apparatus for measuring the extent of partial reflection.

SOURCE: Richard P. Feynman, *QED: The Strange Theory of Light and Matter* (Princeton, NJ: Princeton University Press, 2006), 20.

passing through the glass unperturbed and reaching detector B varies from 100% to 84%.

Probability rules absolutely. We cannot tell whether a particular photon will wind up at detector A; all we can do is indicate the ways a photon might behave. Feynman starts with a glass plate in which there is a 4% chance that a photon will arrive at detector A. He draws an arrow whose "length" represents not the length of the path but its probability, the square of this value: 0.2 times 0.2, or 4%. Since there are two paths by which the photon can reach the detector—from the front or back of the glass, Feynman draws two arrows of the same length: the probability is 4% for both paths.

These two arrows also have a specific direction, one that Feynman determines with an imaginary stopwatch. He starts the watch as soon as a photon leaves the light source and stops it when the photon reaches the detector. At that time, the hand's direction corresponds with the direction of the arrow. But the arrow's direction for the reflection from the back of the glass plate is different from that of the front: the former will take longer to arrive at detector A. In addition, the direction of the front reflection arrow must be opposite to that of the stopwatch hand. The rest is elementary geometry. In figure 2.2 Feynman links the head of the front arrow to the tail of the back one; then the tail of the front arrow to the head of the

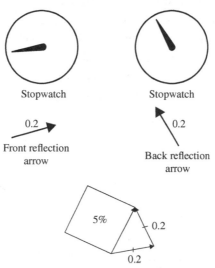

FIGURE 2.2 Percentage of photon reflection for a glass plate with 4% probability that a given photon will reach detector A.

SOURCE: Richard P. Feynman, *QED: The Strange Theory of Light and Matter* (Princeton, NJ: Princeton University Press, 2006), 30.

back one. The square represents the overall probability of reflection from the glass plate: 5%.

Next, Feynman imagines the thickness of the glass plate has increased to the point where the stopwatch hand for the back arrow advances by a half turn, so that the two arrows point in opposite directions. In this case, the probability of a photon reaching detector A has increased to 16%. Analogously, Feynman imagines that the glass plate's thickness has increased to the point where the two arrows point in the same direction. In this case, the probability has decreased to zero.

With gradually thicker layers of glass, the partial reflection of light increases in intensity from near zero to a maximum, decreases to zero, then increases back to the same maximum. The cycle repeats indefinitely. Feynman reports that recent experiments with lasers showed this behavior continues for 100 million repetitions, corresponding to a glass plate having a thickness of 165 feet.

Feynman expands the scope of his explanation to the behavior of light in total reflection. Figure 2.3 represents the classical theory: a light beam from source S strikes a mirror and is detected at P. (The screen Q blocks the beam that would otherwise hit the detector without striking the mirror.) In classical theory, there is only one path from S to the mirror to P, the path of least time, where the angle of incidence equals the angle of reflection.

By sharp contrast, in quantum electrodynamics a photon can travel from S to P by an infinite number of paths, shown in figure 2.4. Each takes a different time to reach its final destination. In addition, because "light has an equal amplitude [probability] to reflect from every part of the mirror, from A to M,"[16] the resulting arrows have different directions. Feynman connects all the arrows, head to tail, forming a long, thick arrow

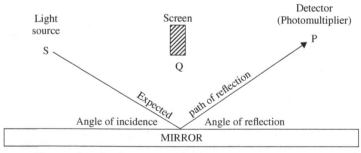

FIGURE 2.3 Picture of total reflection in classical theory.

SOURCE: Richard P. Feynman, *QED: The Strange Theory of Light and Matter* (Princeton, NJ: Princeton University Press, 2006), 39.

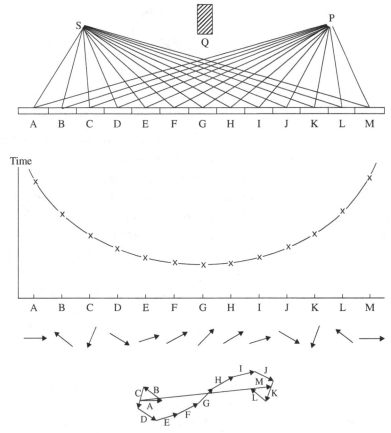

FIGURE 2.4 Picture of total reflection in quantum electrodynamics.

SOURCE: Richard P. Feynman, *QED: The Strange Theory of Light and Matter* (Princeton, NJ: Princeton University Press, 2006), 43.

whose major contributions come from E to I, the paths where the time and distance traveled are least. Feynman concludes:

> To summarize, where the time is least is also where the time for the nearby paths is nearly the same: that's where the little arrows point in nearly the same direction and add up to a substantial length; that's where the probability of a photon reflecting off a mirror is determined. And that's why, in approximation, we can get away with the crude picture of the world that says that that light only goes where *time* is *least*.[17]

At this point, Feynman broadens his explanatory scope to encompass electrons in interaction with photons. In figure 2.5, an example of the famous Feynman diagrams that still do yeoman's service in

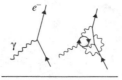

FIGURE 2.5 Diagrams representing electron-photon interactions.

contemporary physics, electrons are straight; photons, wiggly lines, labeled with the Greek letter gamma. Time moves from bottom to top. On the left, an electron absorbs a photon; in its turn, the photon emits an electron. On the right, the electron has a more complicated history. This time the electron emits and absorbs several photons. The circle represents the annihilation of an electron-positron pair issuing in a photon. (A "positron" is an electron with a positive charge.) Whatever the complications in these interactions, there are only three possibilities: a photon goes from place to place; an electron goes from place to place; and an electron emits or absorbs a photon in the process: "It is hard to believe that nearly all the vast apparent variety in Nature results from the monotony of repeatedly combining just these three basic actions. But it does."[18]

Feynman broadens the scope of his explanation still further. Photons and electrons behave differently. Photons can accumulate at a point, increasing the intensity of their effect. This is how lasers work. Electrons behave just as we do: one person to a seat, one electron to a particular spot in spacetime. This exclusion principle accounts for every element that makes up our world, except for hydrogen with its one electron:

> One proton exchanging photons with one electron dancing around it is called a hydrogen atom. Two protons exchanging photons with two electrons [differing in the orientation of their spin] is called a helium atom. You see, the chemists have a complicated way of counting: instead of saying "one, two, three, four, five protons," they say, "hydrogen, helium, lithium, beryllium, boron."[19]

The readership of *QED* is wide indeed; even physicists have been enlightened. Leigh Palmer, an experimentalist at the Simon Fraser University, had without success asked his theoretical colleagues to explain the "Hanbury-Brown, Twiss effect" by means of which it is possible to distinguish one source of interstellar light from another, even when these sources are extremely close. How is such fine discrimination possible? Only when he read *QED* did he discover that the explanation lay in Feynman's analysis of partial reflection:

> While I was following the very lucid development, filling in what I feel is a remarkably small number of "details," and understanding what I was reading, I came across a footnote. It seems that the Hanbury-Brown, Twiss effect had been explained to me entirely *en passant* [a chess move in which a pawn is captured without any action on the part of the opponent]. I suddenly had the rush of understanding, though I did go back and read it again to make sure.[20]

It may seem surprising Feynman explains quantum electrodynamics even to physicists. This is possible because theirs is a worm's-eye view. Working wholly within the quantum world, they turn equations into theories and, through experimentation, theories into facts. Feynman gives them what they do not have, the big picture in lucid prose translation. What is true for the prose is true for the diagrams in *QED*. The general readers' "little arrows" are physicists' "probability amplitudes"; the unnamed depictions of photon-electron interactions are Feynman diagrams.

In *QED*, Feynman recognized a central problem for the general reader: believing that something as bizarre as quantum electrodynamics is true. It is, after all, a theory so defective that its equations need an apparently arbitrary adjustment called renormalization, "a dippy process" necessary to avoid stubborn nonsensical results.[21] Nevertheless quantum electrodynamics is, Feynman believes, a theory to be proud of: its agreement with experiment is nothing less than astonishing. The measurement of the magnetic moment of the electron, an effect of its spin and electrical charge, is 1.00115965218073 (±28); the best theoretical value is a very close 1.00115965218113 (±86). This is the equivalent "of measuring the distance from the Earth to the Moon to within the width of a single human hair."[22]

In his explorations of the micro-world, Feynman is the descendent of the Edmund Burke, who, reflecting on the tiniest of animalcules that 18th-century microscopes revealed, observed that "we become amazed and confounded at the wonders of minuteness; nor can we distinguish in its effect this extreme of littleness from the vast itself. For division must be infinite as well as addition; because the idea of a perfect unity can no more be arrived at, than that of a complete whole, to which nothing may be added."[23] Feynman's is a minuteness Burke could not have imagined, a world where division has reached its apparent limit, a world where the mathematical and dynamic sublime prevails at its lowest possible level. We shall see that if Brian Greene's string theory turns out to be true (chapter 5), it is far from the very bottom.

The Character of Physical Law

In *The Character of Physical Law* Richard Feynman has crafted a masterwork of a different kind, a set of philosophical ruminations on the scope and limitations of physics, a discourse on how mathematics acts as an open sesame to the scientific sublime, the laws that determine the shape and substance of the world we inhabit. Feynman begins with that most famous of laws, Newton's formula for gravitational force. According to this formula, $F = G \, mm'/r^2$, the gravitational force between two objects is the product of their masses divided by the square of the distance between them. This total is then multiplied by G, the gravitational constant.

From this apparently modest beginning, Feynman moves outward to the universal scope of gravity. The force that keeps us glued to our seats is the same force that keeps the planets in their orbits, the same that forms stars from accumulations of gases, the same that collapses those stars into black holes. Gravity's reach extends to the edges of our galaxy; it extends beyond to the universe as a whole. In this biggest of pictures, the earth's gravity

> becomes weaker and weaker inversely as the square of the distance, divided by four each time you get twice as far away, until it is lost in the confusion of the stronger fields of other stars. Together with the stars in the neighborhood [gravity] pulls the other stars to form the galaxy, and all together they pull on other galaxies and make a pattern, a cluster of galaxies. So the earth's gravitational field never ends, but peters out very slowly in a precise and careful law, probably to the edges of the Universe.[24]

$F = G \, mm'/r^2$ is emblematic of modern physics, a discipline mathematical at its core, a research program for which Newton's *Principia* forms the initial and long-enduring foundation.

For Feynman, Newton's proof that Kepler's second law is a consequence of that formula is exemplary of the way theoretical physics works. Kepler discovered that in an elliptical orbit equal times sweep out equal areas. To prove this, we need geometry; we need to show that the area triangle S12 in figure 2.6 equals that of triangle S24. In the figure, the

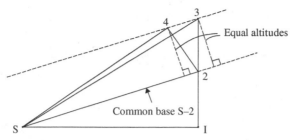

FIGURE 2.6 Geometric proof of Kepler's second law. Based on figure in Newton's *Principia*.

SOURCE: Richard P. Feynman, *The Character of Physical Law* (Cambridge, MA: MIT Press, 1965), 43.

number 3 is the point to which inertia would have taken the planet in the absence of the sun's gravitational force; the number 4 is the actual location of the planet due to that force. Feynman starts by proving that S12 = S23, then that S23 = S24. As Feynman quips, "Isn't that ingenious? I borrowed it straight from Newton. It comes right out of the *Principia*, diagram and all. Only the letters are different because he wrote in Latin and these are Arabic numerals."[25]

While physics is mathematical, it is not only mathematics. The test of the truth of theories is their fit with the world, a fit only experiment reveals. It is experimentation that turns conjectural science into consensual science. Newton's equation $F = G \, mm'/r^2$ cannot be solved unless the constant G, the gravitational constant, is determined experimentally. As Feynman explains, this can be accomplished by a version of an experimental apparatus first employed by the great English experimentalist Henry Cavendish. It is re-imagined in figure 2.7.

In this experiment, the smaller spheres are attracted to the larger; as a consequence, the vertical wire twists clockwise. By measuring the torsion in this wire, scientists can determine the degree of mutual attraction between the spheres. The smaller sphere is then weighed, giving us the force between it and the known mass of the earth. It is now possible to determine G in Newton's formula for the force of gravity.

Feynman makes another point, one about the discovery of laws like Newton's. It is fortunate for the future of physics that more than one mathematical formulation can address the same set of problems. It is a flexibility that enables discovery. Newton's law addresses gravitation, but so does Einstein's general relativity, a field theory, the application of another sort of mathematics, Riemannian geometry. While these two formulations are mathematically equivalent in most applications, "psychologically they are

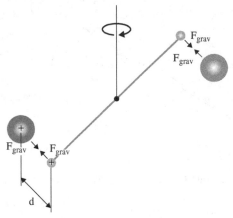

FIGURE 2.7 Cavendish's torsion balance.

different because they are completely unequivalent when you are trying to guess new laws."[26] Since Newton's theory implied that the force of gravity was instantaneous, Einstein realized that it must be wrong: no particle or force field travels faster than the speed of light: "So in Einstein's generalization of gravitation Newton's method of describing physics is hopelessly inadequate and enormously complicated, whereas the field theory is neat and simple."[27] For physicists, neat and simple always trumps enormously complicated.

Feynman outlines two other paths to the scientific sublime in physics, the true character of elementary particles and forces. The first is analogy. Some things, like electrical charge and energy, are conserved, that is, no matter what happens, there is never any less of either of them. Unlike energy, however, charge is not only conserved; it comes in countable units like one plus charge or two minus charges. It is also the source of fields, regions in which each point is affected by a force. Like electrical charge, elementary particles called "baryons" are also conserved. But does the analogy hold between them and charge? Do they also come in units? They do. Are they also the source of a field? Perhaps. "If charge is the source of a field, and baryon does the same things in other respects it ought to be the source of a field too. Too bad that so far, it does not seem to be; it is possible, but we do not know enough to be sure."[28]

Symmetry is another source of revelation, of new laws in physics. Translation in space is one sort of symmetry: your Cuisinart works as well in the living room as in the kitchen. Another type of symmetry is a legacy of special relativity. According to the principle of symmetry for uniform motion, there is no "now" that two different observers of the same event can share. This has an interesting and unexpected consequence: "The clocks inside [a] space ship are ticking at a different speed (slower) from those on the ground."[29] Feynman applies this principle to the mu meson, a subatomic particle that is created as a result of cosmic rays in the outer atmosphere and has a lifetime of only two millionths of a second. Despite this brief lifetime, in a cosmic ray shower, mu mesons easily reach the earth because they are moving at close to the speed of light and time for them has slowed considerably relative to us.

While Feynman celebrates science at every turn, he also recognizes four limitations to its power. First, although the laws of science can be very accurate, they cannot be exact. Einstein's law of gravity was an improvement in accuracy over Newton's. But there is a limit. Moreover, in the underlying micro-world probability rules, not certainty. Can we measure both the position and velocity of an electron with accuracy? We cannot. In the quantum world, "there is always an edge of mystery, always a place where we have some fiddling to do yet."[30]

There is a second limitation. Our ability to pinpoint certain aspects of the world does not imply that we can discover the mechanism behind them, that we can look under the hood. When his critics complained to Newton that his theory of gravity didn't tell them what gravity was, he might have said that it tells you how the planets move, not why. This is still true:

Up to today, from the time of Newton, no one has invented another theoretical description of the mathematical machinery behind this law which does not either say the same thing over again, or make the mathematics harder, or predict some wrong phenomena. So there is no model of the theory of gravitation today, other than the mathematical form.[31]

This is doubly true of quantum mechanics. Feynman admonishes us not to ask how the world can be like that; nobody knows: "Nobody understands quantum mechanics."[32]

There is a third limitation. Although the laws of science can predict the next arrival of Halley's comet, they cannot predict the winner of the next World Series. Nor can the past be changed. The Battle of Hastings was fought on October 14, 1066; it will have been fought on the date forever. Our future, however, seems open to us. There is the sense that we exist on the advancing edge of a perpetual now. But it is an instant whose meaning science cannot plumb. Nor can science help us understand time as we experience it. Science tells us that a universal clock exists: entropy, the increasing disorder in the cosmos. But we do not experience entropy; we experience instead its apparent opposite: automobiles are built, babies are born. It does not help to know that we can "add to the physical laws the hypothesis that in the past the universe was more ordered, in the technical sense, than it is today."[33]

There is a final limitation. As we move up the ladder of complexity from electrons to frogs to history, as we explore such central concepts as evil and beauty and hope, the laws of science become increasingly irrelevant. It would be nice to be able to join science and the world we experience into a significant whole, an all-encompassing theory of everything. But, Feynman avers, we can't do that—at least not yet. In the meantime, it would fool-ish to start a war with the humanities, the custodians of history and hope. Neither set of disciplines is "nearer to God if I may use a religious meta-phor."[34] Nevertheless, there is, he feels, a movement toward convergence:

> The great mass of workers in between [fundamental laws of science and fundamental cultural ideas], connecting one step to another, are improving all the time our understanding of the world, both from working at the ends and working in the middle, and in that way we are gradually understanding this tremendous world of interconnected hierarchies.[35]

The Character of a Scientist

Feynman became an important figure in the scientific community for numerous reasons: his work at Los Alamos during World War II, his *Lectures on Physics*, and his contributions to quantum electrodynamics leading to a Nobel Prize in 1965. But it was not until the 1980s and the publication of *Surely You're Joking, Mr. Feynman!* and *What Do You Care What Other People Think?* that he truly entered the ranks of popular scien-tists. In these books, we see the scientist as a raconteur, a practical problem solver, a celebrity, an embodiment of the scientific sublime.

On February 15, 1988, Richard Feynman died. At the memorial serv-ice, Murray Gell-Mann angered his family by saying that the physicist had "spent a great deal of time and energy generating anecdotes about himself," stories "in which he had to come out, if possible, looking smarter than any-one else."[36] James Gleick, his otherwise sympathetic biographer, agrees. He feels that Feynman's "own view of himself worked less to illuminate than to hide the nature of his genius."[37] Nothing could be further from the truth. *Surely You're Joking, Mr. Feynman!* is not just entertainment, though entertainment it is. It also tells us what it is to do theoretical physics like Richard Feynman, physics by a fox who works alone, enamored since boyhood by difficult puzzles, and confident since boyhood that their solution is just around the corner.

The Scientist as Raconteur

We first meet Feynman as a boy with an unusual hobby: he fixes radios for pocket change. Adults willingly employ him: after all, there is a Depression on; no one has much money. At first, Feynman knows nothing about radios. But his learning is not book learning; he consults no manuals; he observes no repairmen. He just tackles each problem head-on, absolutely certain that its solution is near at hand. One broken radio is a particular challenge; it gives out an initial blast of sound, a hellish screech. The solution is far from obvious, so Master Feynman paces back and forth, thinking. It is a maneuver that makes the radio's owner nervous: walking back and forth is not radio repair. But the guy is wrong—Feynman now has a theory worth testing: the radio's tubes have been inserted in the wrong order. Sure enough, when the order is reversed, the radio behaves properly. The man is stunned and appreciative of this early experimental proof:

> He got me other jobs, and kept telling everybody what a tremendous genius I was, saying, "He fixes radios by *thinking*!" The whole idea of thinking, to fix a radio a little boy stops and thinks, and figures out how to do it—he never thought that was possible![38]

Feynman, now a young Cornell professor, sits in the university cafeteria and solves another puzzle by "thinking." Observing a student idly spinning a plate into the air, he tries to calculate the relationship between its spin and its wobble:

> I went on to work out equations of wobbles. Then I thought about how electron orbits start to move in relativity. Then there's the Dirac Equation in electrodynamics. And then quantum electrodynamics. And before I knew it (it was a very short time) I was "playing"—working, really—with the same old problem that I loved so much; that I had stopped working on when I went to Los Alamos: my thesis-type problems; all those old-fashioned, wonderful things. It was effortless. It was easy to play with these things. It was like uncorking a bottle. Everything flowed out effortlessly. I almost

tried to resist it! There was no importance to what I was doing, but ulti-mately there was. The diagrams and the whole business that got the Nobel Prize for me came from that piddling around with the wobbling plate.[39]

It is this way of viewing the world that played a crucial part in a landmark paper Feynman wrote with his colleague Murray Gell-Mann. He is chat-ting with some fellow physicists about physics:

> Finally they get all this stuff into me and they say, "The situation is so mixed up that even some of the things they've established for *years* are being questioned—such as the beta decay of the neutrons is S and T. Murray says it might even be V and A, it's so mixed up."
>
> I jump up from the stool and say. "Then I understand EVVVVVERYTHING."
>
> They thought I was joking. But the thing I had trouble with at the Rochester meeting—the neutron and the proton disintegration: everything fit *but* that, and if it was V and A instead of S and T, *that* would fit too. Therefore I had the whole theory! . . . It was the first time that I knew a law of nature that nobody else knew.[40]

Feynman and Gell-Mann were so confident that their theory was correct that they dismissed as mistaken experiments that apparently undermined it: "These theoretical arguments seem to the authors to be strong enough to suggest that the disagreement with the He^6 recoil experiment and with some other less accurate experiments indicates that these experiments are wrong."[41] Elsewhere Feynman enunciates the general principle: "We've learned from experience that the truth will come out. Other experimenters will repeat your experiment and find out whether you were wrong or right. Nature's phenomena will agree or they'll disagree with your theory."[42] The consensus of scientists is a coincidence between their work and the world.

The Scientist as Practical Problem Solver

On June 9, 1986, President Reagan received from the Rogers Commission a report on the causes of the *Challenger* disaster. It contained a separate section, Appendix F, written by Richard Feynman. It was his last public act. Desperately ill from the recurrence of a rare cancer, he had risen to this challenge as a call to duty motivated by a deep-seated emotional commit-ment to science. It was also motivated by the knowledge that he was the only member of the Commission who would be savvy enough to discover the mechanical cause of the disaster and independent enough to tell the truth to the American public. He was, it turned out, not savvy enough: the mechanical cause, the O-ring's lack of resilience in cold weather, eluded him. This was hardly surprising. The disaster could have been caused by any of number of problems with this elaborate machine. Anything could go fatally wrong; something would go fatally wrong, given a sufficient number of flights. For example, the *Columbia* disaster nearly two decades later was caused by a large piece of foam that fell from the shuttle's external tank and breached the spacecraft wing.

The mechanical cause was not, of course, the real cause. The real cause was the culture of NASA, a heady brew of unfounded optimism. NASA estimated the flight risks as 1 in 100,000. Feynman retorted: "That means you could fly the shuttle *every day* for an average of *300 years* between accidents—every day, one flight, for 300 years—which is obviously crazy."[43] NASA engineers guessed the number as 1 in 200 or 1 in 300.[44] In fact the actual number is nearer 1 in 100. There have been around 200 flights, two disasters, 13 dead astronauts, and one dead passenger, Christa McAuliffe.

Caught in the swirl of Washington politics, bureaucratic obfuscation, and engineering confusion, Feynman was befriended by Major General Donald Kutyna, who nudged him in the direction of the crucial mechanical defect. Kutyna tells Feynman an apparently irrelevant anecdote:

> I was working on my carburetor this morning, and I was thinking: the shuttle took off when the temperature was 28 or 29 degrees. The coldest temperature previous to that was 53 degrees. You're a professor; *what, sir, is the effect of cold on the O-rings?*[45]

With this crucial inference in hand, Feynman got into gear. Was it really true that the O-rings would stiffen with cold, break the seal, and cause the disaster? A physics experiment was the answer. While Feynman realized that it would be "more honest and dramatic to do the experiment for the first time in the public meeting of the Presidential Commission,"[46] he could not resist trying his luck ahead of time, just to make sure.

At the public meeting, Feynman sat impatiently with an O-ring segment, a glass of ice water, and a C-clamp:

> General Kutyna, who's caught on to what I'm doing, quickly leans over to me and says, "Co-pilot to pilot, not now."
>
> Pretty soon, I'm reaching for my microphone again.

"Not now!" He points in our briefing book—with all the charts and slides Mr. Mulloy is going through—and says, "When he comes to this slide, here, that's the right time to do it."[47]

The right time is when all the television cameras were pointed in Feynman's direction. That night this feat of showmanship was on all the networks; the next morning Feynman was on the front page of the *New York Times*. On June 9, the day the President received the report, he was interviewed on *The MacNeil/Lehrer News Hour*.[48]

Edward Tufte faults Feynman. His was not really an experiment: there was no control. For that you need two glasses of water—one with and one without ice. You need this to show that the ice alone is the culprit.[49] But Feynman was not experimenting, despite what he said; he was revealing to the Commission and to the American people what he knew a real experiment would bear out: the findings of consensual science, an experiment that would in fact put the general public in touch with the scientific sublime. That day he was an inspired teacher dramatizing the truth, not only in newspapers and on television but also in more permanent form, in a book meant for a wider public and for future generations: *What Do You Care What Other People Think?* It is still selling well, a quarter century after its publication. In this last public appearance, although Feynman deviated from experimental protocol, he did not deviate from the truth. Of the O-rings he wrote that they "were not designed to erode. Erosion was a clue that something was wrong. Erosion was not something from which safety could be inferred."[50]

The Scientist as Celebrity

Hardly anyone turns down the Nobel Prize, a prize for science that has become, resoundingly, consensual; no scientist ever has. It is, one assumes, a pleasure to be chosen as singularly outstanding, an embodiment of the scientific sublime. So it may seem that Richard Feynman was being insincere when he agreed to accept the Prize reluctantly. But his reluctance is entirely in character. A letter resigning from the National Academy of Science explains his discomfort with public honors:

> To be a member of a group of which an important activity is to choose others deemed worthy of membership in that self-esteemed group, bothers me. The care with which we select "those worthy of the honor" of joining the

Academy feels to me like a form of self-praise. How can we say only the best must be allowed in to join those who are already in, without loudly proclaiming to our inner selves that we who are in must be very good indeed. Of course I believe I am very good indeed, but that is a private matter.[51]

From that long-ago meeting at which he bumped into Murray Slotnick, he knew he was very good indeed. Fame was as incidental as dressing up for dinner. This distaste for public pomp and circumstance is deeply embedded in Feynman's character. Reminiscing about his relationship with his father, he reiterates a persistent theme in Melville Feynman's conversation: the uniform is not the man. The pope, his father says, is just a human being in fancy dress: "The difference was the hat he's wearing."[52]

For Feynman, fame could become downright annoying. After receiving the Prize, he gives a public talk on a technical subject. To his dismay, he finds his audience has swelled with people for whom the talk will be incomprehensible. He does not have an audience; he has fans, the status that Stephen Hawking enjoys Richard Feynman deplores. Forewarned is forearmed. Invited to speak at a college science club, he will accept only on the condition that the talk is advertised in exact accord with his wishes. The speaker on the fliers announcing the talk will be a fictitious nonentity; his topic will be guaranteed to bore. Feynman's ruse succeeds. He arrives and apologizes; the advertised speaker could not make it; Feynman has been called in as his last-minute substitute.[53]

Feynman knew the difference between reputation and fame. He cherished the first and dismissed the second. His Banquet Speech at the Nobel Prize ceremonies makes this clear:

The work I have done has, already, been adequately rewarded and recognized. Imagination reaches out repeatedly trying to achieve some higher level of understanding, until suddenly I find myself momentarily alone before one new corner of nature's pattern of beauty and true majesty revealed. That was my reward. Then, having fashioned tools to make access easier to the new level, I see these tools used by other men straining their imaginations against further mysteries beyond. There are my votes of recognition.[54]

At one with the scientific sublime, Feynman is eager to share his experience with others—other physicists of course, but the general public as well. Of him we may say what Geoffrey Chaucer said of his Canterbury scholar: "Gladly would he learn and gladly teach."

Conclusion

Feynman confines himself to what physicists actually know, physical laws backed by sound experimentation. He begins with the origin of modern physics in the 17th century and stops with mid-20th-century quantum electrodynamics, a theory firmly supported by experimental data. His is a consensual sublime, conveyed by the interaction of plain English with simple diagrams. In his autobiographical writings, he recounts the origin of a man devoted to the pursuit of this consensual sublime and to communicating it to others. Throughout all his books, the written word captures his unique and engaging voice, whether he is explaining science or recounting an anecdote. This is not surprising: both *The Character of Physical Law* and *QED* have lectures as their origin, and the two autobiographical volumes are the product of taped and edited interviews. Those who have not had the pleasure of seeing Feynman the lecturer and raconteur in action should check out the many YouTube videos that are freely available.

CHAPTER 3 | Steven Weinberg
The Conjectural Sublime

The more the universe seems comprehensible, the more it also seems pointless.

—STEVEN WEINBERG

STEVEN WEINBERG HAD been working for some time on the problem of the strong force that holds together the components of an atom's nucleus. He was getting nowhere. "Suddenly," while driving home in his red Camaro, insight arrived. He did not have the wrong answer to the problem of the strong force but the right answer to a different, equally interesting problem:

> And I realized the massless particle in this theory that had given me so much trouble had nothing to do with the heavy particles that feel the strong inter-action; it was the photon, the particle of which light was composed, that is responsible for electric and magnetic forces and that indeed has zero mass. I realized that what I had cooked up was an approach not just to understand-ing the weak interaction but to unifying the theories of the weak and elec-tromagnetic forces into what has since come to be called the electroweak theory.[1]

"A Model of Leptons"[2] is a paper of which he is justly proud. It has gar-nered 4,503 citations; a copy has been offered for sale at $950. This is the physicist at his mathematical best, a language he speaks as if it were his native tongue.

Another incident confirms Weinberg's extraordinary talent. Physicist Rich Muller has a bright idea. After several tries, however, the mathematics continues to defeat him. Despondent, he walks down the hall to an office where Steven Weinberg is chatting with Freeman Dyson. The two agree to help:

> Weinberg went to the blackboard, wrote down the first equation. "And then he did some manipulations on it," said Muller, "and stood back." Dyson said, "I think if you make a substitution of variables now—." Weinberg said, "Oh, yes, of course," and wrote several more lines. "I was taking notes," Muller said, "but I wasn't sure what he was doing." Weinberg paused in his writing, and Dyson said, "Now evaluate the delta function," and Weinberg said, "Oh, okay." Weinberg wrote down a few more lines, and Dyson said, "Good. You've proven it." Muller's idea was right. Weinberg and Dyson, said Muller, "are two of the most incredibly brilliant people I have ever met."[3]

It is an anecdote that Richard Feynman would have loved to tell about himself. But Weinberg is a private person; he does not talk about himself or his personal life. He met his wife, Louise, a law professor at the University of Texas, when they were Cornell undergraduates. They have one daughter, Elizabeth. He once owned a red Camaro. That is all we know. No *Surely You're Joking, Mr. Weinberg* will be forthcoming.

A Nobel Prize–winning physicist, Weinberg has also had a second career as a popularizer of physics. In all his popularizations, he shows us how much we can learn about physics with only a little arithmetic, and how much more when a little algebra is thrown in, along with some equations, carefully segregated in extensive endnotes. The science he describes conveys not only the consensual sublime favored by Feynman but also the conjectural sublime, summoning up for us what might have occurred in the first three minutes of the universe and what shape a final theory might one day take, based on the best available experimental and observational evidence.

The First Three Minutes

The old-time radio show *You Are There* traveled backward in time to the assassination of Lincoln, the defeat of the Spanish Armada, and the execution of Emperor Maximillian. In his first and most successful popular science book, *The First Three Minutes: A Modern View of the Origin of the Universe*, Steven Weinberg engages in a different kind of time travel, one that takes us back to a time when nobody was there, when nobody could be there. While the evidence for assassinations, defeats, and executions lies in the archives, the evidence for the beginning of the universe surrounds us, although it observable only with highly specialized scientific instruments, observations seen through the lens of abstruse theory.

To see our world for what it was, we must discard the deepest illusion we harbor—its stability. To penetrate this illusion is to experience two alien worlds—the incredibly vast and the incredibly minute. Although both hold the key to the evolution of the universe, it is the micro-world that is the more fundamental to our understanding. It is a world of elementary particles ruled by the laws of quantum mechanics, a universe that began fourteen billion years ago in an explosion, one that, unlike ordinary explosions, took place everywhere as space itself expanded. We know this Big Bang took place, though there was no time at which and no space in which it could take place. Neither time nor space existed at $t = 0$. Quantum mechanics is physics through the looking glass, a series of bold conjectures from which a conjectural past can be constructed.

Of all our popular science physicists, Weinberg is the most far ranging in his literary references. It comes as no surprise that this frequent contributor to the *New York Review of Books* begins his cosmic origin story with the Norse Eddas. According to this myth, the universe arose from a state of nothingness. Out of regions beyond nothingness, fire melted frost and produced godlike giants and a cow to nourish them. Although Weinberg is dismissive of this myth's absurdities, he is hardly dismissive of the problem these are meant to explain: "From the start of modern science in the sixteenth and seventeenth centuries, physicists and astronomers have returned again and again to the problem of the origin of the universe."[4] While his scientific account of the universe's beginning does not match the inscrutable nobility of Genesis, it is equally an invocation of the sublime, an account of high drama among elementary particles. Weinberg's relation of our conjectural end also matches in frisson the apocalypse of Revelations. His story, however, is not like the Bible, founded on faith; it must be argued into place, inferred from the evidence science and only science can muster.

Looking out at the night sky, we see the stars fixed in place, the moon in its monthly phases, a permanent clockwork with us as its center. Only science can reveal a very different picture, a universe that began with a big bang about fourteen billion years ago and has been expanding ever since, a cosmos with no center at all. It is up to *The First Three Minutes* to show us that this unlikely picture is a fact no sensible person can deny. But there is a problem in addressing general readers, the difficulty of conveying the revelations of a discipline as intensely mathematical as theoretical physics.

Before beginning his account, therefore, Weinberg assures us that he pictures "the [typical] reader as a smart old attorney" innocent of advanced mathematics, but one who "expects nonetheless to hear some convincing arguments before he makes up his mind."[5]

One of these arguments hinges on the Doppler effect. Even in the absence of partial differential equations, even in a work that relies solely on the knowledge of arithmetic and elementary algebra, readers are faced with the intellectual challenge of the complex physics behind the Doppler effect. Weinberg meets it with expository brilliance: readers can follow his narrative in three modalities. First, they can grasp it in his straightforward prose exposition. If they fail to do so, they have another chance: they can grasp it by means of Weinberg's analogies with the everyday. If they can follow Weinberg's algebra, however, they can understand the Doppler effect in its proper language. For those readers, the primary textual exposition and the analogy merely assist in understanding the algebraic formulation. And it is only those readers who touch base with the power of mathematics in physics; it is only they who achieve the deepest level of understanding. In the words of another Nobel laureate, Eugene Wigner, these readers share a daily miracle in which all physicists participate: "The miracle of the appropriateness of mathematics for the formulation of the laws of physics is a wonderful gift which we neither understand nor deserve."[6]

According to Weinberg, the Doppler effect occurs because "a wave sent out by a source moving away from us will appear to have a *longer* wavelength than if the source were at rest. . . . Similarly, if the source is moving toward us, the time between arrivals of wave crests is decreased because each successive crest has a shorter distance to go, and the wave appears to have a *shorter* wavelength."[7] For those who might find this formulation opaque, Weinberg tries an analogy:

> It is just as if a traveling salesman were to send a letter home regularly once a week during his travels: while he is traveling away from home, each successive letter will have a little farther to go than the one before, so his letters will arrive a little more than a week apart; on the homeward leg of his journey, each successive letter will have a shorter distance to travel, so they will arrive more frequently than once a week.[8]

The effect is easily demonstrated. You hear the high pitch of the siren of the approaching ambulance; you notice that its pitch drops as the ambulance departs. In the case of light, the effect is more complex. By analogy,

Doppler thought that he could determine whether a star was moving away or toward the earth by its color: the stars moving away from earth would have a longer wavelength and be redder than average; stars moving toward earth would have shorter wavelength and be bluer. He was wrong: the color of stars depends largely on their surface temperature. But there really is a red shift in the case of stars. As discovered by William Huggins in 1868, it is dark lines within the light spectrum that shift toward the blue or red for approaching or receding stars. For those for whom elementary algebra is not a foreign country, in an endnote Weinberg derives the mathematical relationship between the length of wave crests emitted by the light of moving astronomical objects and their velocity.[9] From that relationship, readers who so desire can calculate the velocity of an astronomical object from its redshift.

The Doppler effect was the clue that enabled Edwin Hubble to conjecture that the universe might be rapidly expanding, a story that turns Weinberg from a scientist to a historian of science. In 1929 Hubble analyzed the available observational data from 46 galaxies and came to a surprising conclusion: they do not remain stationary or move around randomly; the farther away a galaxy is, the faster it appears to be moving away from us. Hubble's evidence for this relationship is communicated in a line graph plotting velocities calculated from measured redshifts versus the mean distance estimated from apparent luminosities of the galaxies.

On Hubble's graph with its relatively few data and large scatter, Weinberg in characteristically blunt: "A look at Hubble's data leaves me perplexed how he could reach such a conclusion—galactic velocities seem almost uncorrelated with their distance, with only a mild tendency for velocity to increase with distance. . . . It is difficult to avoid the conclusion that . . . Hubble knew the answer he wanted to get." Perhaps. But I would note that in his published paper Hubble shows extreme caution. He characterizes the linear velocity-distance relationship as a conjecture, "a first approximation representing a restricted range in distance." What's more, he refuses to speculate on the most obvious interpretation of its significance: "New data to be expected in the near future may modify the significance of the present investigation or, if confirmatory, will lead to a solution having many times the weight. For this reason it is thought premature to discuss in detail the obvious consequences of the present results."[10]

Even in the 1930s, when data poured in supporting his "first approximation" Hubble resisted the explanation of an expanding universe. He viewed it as only one of several plausible explanations consistent with the evidence. Even when his results were confirmed, he persisted in regarding the relationship he discovered as stubbornly empirical. Hubble was correct in his caution. His work and that of his successors was suggestive rather than probative—a conjecture, not a fact. On that, Weinberg heartily agrees; he makes it clear that from this evidence it only "*appears* that the universe is undergoing some sort of explosion in which every galaxy is rushing away from every other galaxy."[11]

Only with the discovery of cosmic microwave radiation was the case for an expanding universe clinched, conjecture turned into fact. In theoretical papers beginning in 1948, Ralph Alpher and Robert Herman predicted the existence of this background radiation permeating the universe. It should have, they reported, a temperature of 5 kelvins, 5 degrees Celsius above absolute zero. It was not, however, until close to two decades later that anyone confirmed that prediction. Weinberg does a masterful job of accounting for the puzzling delay. First, Alpher and Herman were using the wrong Big Bang model, one that could not account for the formation of the heavy elements; second, most physicists were under the impression that cosmic microwave radiation could not be detected; third, most did not think that the origin of the universe was a well-formed problem, capable of solution with the theoretical and observational tools on hand. Weinberg might have added that Alpher and Herman did not follow up on their initial paper; he might also have added that, having become industrial physicists, they were outside the Ivy League–Chicago-Texas-California–Bell Labs charmed circle.

In 1965, Weinberg informs us, radio astronomers Arno Penzias and Robert Wilson published the significant observational result: a cosmic radiation measurement of 3 degrees above absolute zero, in reasonable agreement with Alpher and Herman's earlier calculation. In an accompanying paper, Robert Dicke, Jim Peebles, Peter Roll, and David Wilkinson concluded that Penzias and Wilson had measured the residue of the explosion with which the universe began.[12,13]

There is a second conundrum Weinberg resolves, turning conjecture into fact. It concerns cosmology not as a science but as a social system. Why, given Alpher and Herman's clear theoretical priority, did Penzias and Wilson win the Nobel Prize? In 1985 physicist Martin Harwit asks Robert Dicke whether he thinks it "amusing" that Penzias and Wilson won the Prize for something they stumbled on, something whose significance they did not understand. Dicke opines that "one would think that pure serendipity shouldn't be so strongly rewarded." When Harwit suggests that the award might have been given to Alpher and Herman, Dicke demurs: "I would have found it objectionable." It is Harwit who makes the essential point:

> You know, to some extent the physics community has tended to pride itself on valuing an incisive theoretical prediction, more highly often than even the measurements which then verify that prediction. And when the measurements are not only made by chance; but have to be explained to the experimenters by somebody else, it seems to me that the choice between those two might be quite clear-cut.[14]

Dicke responds: "Okay, I think I'm rather almost inclined to agree with you on this, that in one case you're rewarding pure serendipity."

For there is no doubt whatever that Penzias and Wilson stumbled on their result. The early 1960s had seen the beginning of transatlantic television, powered by a transmission system in which signals bounced off a balloon. This signal was so weak, however, that the problem arose of dealing with interfering signals. This is the problem Penzias and Wilson addressed; cosmic microwave radiation was the last thing on their minds. But because TV signals are microwave signals, they pointed an old microwave antenna at the sky to see if anything interesting turned up. Then they ran into a persistent problem they could not solve: no matter how they oriented their antenna, a faint signal came from every direction. At first, they suspected a flaw in their equipment; this turned out not to be the case. It was at this point that they learned that some Princeton astrophysicists might be able to turn their abject failure into a resounding success. As their paper clearly reveals, even after conversations with these astrophysicists, they were not willing to endorse the real significance of their discovery.[15] In their interviews with Harwit, Alpher and Herman are understandably bitter at the failure of the physics community to acknowledge their important contribution properly.

Weinberg is a notable exception to this persistence of this injustice. He sends them relevant portions of his original draft of *The First Three Minutes*, an effort that, understandably, elides the importance of their contribution. Having read the original papers at their urging, he acknowledges that their work "was, in fact, the first thoroughly modern analysis of the early history of the universe."[16] It cannot be very often that an amateur historian writing a book of popular science sets the record straight on so important a matter. Of course, it may seem that the matter is not important at all, that the neglect of Alpher and Herman can itself be safely neglected because it is irrelevant to the progress of cosmology. This would be a mistake. The ability of cosmology to reveal the patterns nature so deceitfully hides and to infer from these the laws she so servilely obeys depends crucially on the justness of the system of rewards on which the reputations of scientists rest. It's all right to ignore the literature in your search for new ideas: this was Feynman's way and it did him proud. But it's not all right to ignore the literature when you publish your results or distribute awards. It is a tribute to Weinberg that he worked to correct an injustice; it is a pity that it was necessary for him to do so.

With Penzias and Wilson's result properly interpreted, it was now possible for Weinberg to reconstruct the history of the early universe—a *2001* that out-Kubricks Kubrick. In Weinberg's scenario, a movie plot unfolds. Its characters are elementary particles, their behavior motivated by the degree of heat the universe is experiencing. The universe evolves, frame by frame:

> FIFTH FRAME. The temperature of the universe is now 1,000 million degrees Kelvin (10^9 K), only about 70 times hotter than the center of the sun. Since the first frame, three minutes and two seconds have elapsed. The electrons and positrons have mostly disappeared, and the chief constituents of the universe are now photons, neutrinos, and anti-neutrinos. The energy released in the electron-positron annihilation has given the photons a temperature 35 percent higher than that of the neutrinos.[17]

Readers should note Weinberg's wonderful "only," a word designed to emphasize how cool the universe has gotten in so short a time, down from a high of 100,000 million degrees kelvin at t = 1/100 second, and how hot it still is, too hot to form the elements that are the subject matter of chemistry. Those will not appear for another 700,000 years.

Weinberg reflects on how well his account of the beginning of the universe will hold up over time. That is, he feels, the wrong question:

> In following this account of the first three minutes, the reader may feel that he can detect a note of scientific overconfidence. He might be right. However, I do not believe that scientific progress is always best advanced by keeping an altogether open mind. It is often necessary to forget one's doubts and to follow the consequences of one's assumptions wherever they may lead—the great thing is not to be free of theoretical prejudices, but to have the right theoretical prejudices. And always, the test of any theoretical preconception is in where it leads. The standard model of the early universe has scored some successes, and it provides a coherent theoretical framework for

future experimental programs. This does not mean that it is true, but it does mean that it deserves to be taken seriously.[18]

It is a paradox of science that the right wrong answers can be very fruitful indeed; such conjectures can be an open sesame to Kant's mathematical and dynamic sublime. Just ask the man in the red Camaro.

The Subatomic Particles

According to *The First Three Minutes,* the universe began as an unimaginably small, dense, and hot soup of subatomic particles. The first and by far the longest chapter of Weinberg's next book, *The Discovery of Subatomic Particles*, is devoted to J. J. Thomson's 1897 discovery of one of these particles, the electron, the result of an experiment that turned the conjectural into the real. This was the discovery that concluded the era that began with Newton's *Principia*; it marked the beginning of the quantum era, initiated by Thomson's contemporary Max Planck and carried to fruition by such later luminaries as Albert Einstein, Paul Dirac, Niels Bohr, Hans Bethe, and Werner Heisenberg. It was in following the footsteps of these giants that Weinberg earned his Nobel Prize.

Weinberg's history of the electron is the history of our growing understanding of a fundamental phenomenon, electricity, an enterprise that involved the combined efforts of America, England, Denmark, France, and Germany, recruiting scientists of the stature of Benjamin Franklin, Hans Christian Ørsted, André-Marie Ampère, and Heinrich Hertz. It is a story that exemplifies science as an enterprise in which theory and experiment work happily together. In the 18th century, Franklin conjectured that electricity was a single phenomenon; in the 19th century, Thomson concluded that the basis of this single phenomenon is a single, minute negatively charged particle: the electron. Franklin's was a triumph of theoretical unification, Thomson's a triumph of experimentation, a decisive vote favoring one theory over another.

In making clear to the general reader the gist of Thomson's discovery, Weinberg faces a steep expository problem. Because Thomson's experiment incorporates Newton's laws of motion, the laws of magnetism, and the laws of energy conservation, it cannot be fully explained to a nonscientist without explaining the physics of these laws. In fact, Weinberg has chosen the discovery of the electron precisely because it enables him to reveal the way these theories are interwoven into a fine-spun net that, eventually, captures the electron. While these theories cannot be understood without recourse to mathematics—thereby violating a supposed cardinal rule of popular science writing—no more than a knowledge of elementary algebra is required. Braving these difficulties, the intrepid reader will attain a deeper understanding of the way physics works, and the way an experimenter of genius labors to attain that understanding. Still, by following Weinberg's main narrative thread, the less mathematically adept can reach a significant level of understanding of Thomson's great experiment.

It is commonly said of talented experimenters that they have golden hands. But an early assistant noted, "J. J. was very awkward with his fingers, and I found it necessary not to encourage him to handle the instruments."[19] Thomson's genius lay not in the execution of experiments but in their conception, in intuiting exactly the right way to put nature on the rack. Cathode rays were the issue Thomson addressed. Because Heinrich Hertz could not get them to deflect by electrified metal plates, he had inferred that they did not consist of charged particles. Thomson did not believe Hertz's result; he felt that the rays would deflect if an experiment were properly conducted. All he need do was evacuate his apparatus with a superior air pump, one that would create a virtual vacuum.

Thomson was correct. In Figure 3.1, cathode rays in his apparatus traveled past focusing coils, and then past deflecting electrical plates. From there they sped past a deflecting electric or magnetic field. Finally, they entered the drift region, at the end of which they hit a phosphorescent screen, their deviation indicating a stream of negatively charged particles.

The deflection had not only to occur; it had to be seen to occur: "The phosphorescent patch at the end of the tube was deflected, and the deflection measured by a scale pasted to the end of the tube. As it was necessary to darken the room to see the phosphorescent patch, a needle coated with luminous paint was placed that could be moved up and down the scale; this needle could be seen when the room was darkened, and it was moved until

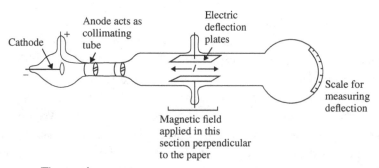

FIGURE 3.1 Thomson's apparatus.

it coincided with the phosphorescent patch. When light was admitted, the deflection of the phosphorescent patch could be measured."[20]

Can we trust Thomson's calculation? Can that result, even if trusted, count as the first discovery of a subatomic particle? Weinberg makes it clear that Thomson's report would never pass muster in today's world of science. Not only was he a clumsy experimenter; he was a careless one. He failed to publish estimates of uncertainty in his measurements; moreover, those measurements were two to three times greater than they should have been. Even when using a method that gave him accurate measurements in two of his three cathode ray tubes, Thomson preferred the inaccurate readings of his third tube. Lastly, the result Thomson found was ambiguous in the inferences it could empower: it could mean either that the mass of the particle was very small or that the charge was very large. Without further evidence as to which alternative was correct, Thomson opted for the former, inferring that the particle mass was much smaller than that of an atom. He leaped (Weinberg's word) to a conclusion to which he was not entitled on the basis of the dubious evidence he brought forward. Thomson asserted that

> on this view we have in the cathode rays matter in a new state, a state in which the subdivision of matter is carried very much further than in the ordinary gaseous state: a state in which all matter—that is, matter derived from different sources such as hydrogen, oxygen, &c.—is of one and the same kind; this matter being the substance from which all the chemical elements are built up. [21]

Thomson's "discovery" of the electron was the consequence of a factor not ordinarily considered science: he was a man given to bold conjecture, a man who felt he had a hotline to the scientific sublime. This is also the opinion of leading physicist George FitzGerald: "I may express a hope that Prof. J. J. Thomson is quite right in his by *no means impossible hypothesis.* It would be the beginning of great advances in science."[22]

Weinberg's is a not entirely successful experiment in giving general readers a taste of what discovery in physics is like. While the intrepid may learn a great deal merely by persisting, the average reader will soon become entangled in an underbrush of barely relevant footnotes, superfluous illustrations, and lengthy digressions. For example, Weinberg's discussion of energy relations is interrupted by a 500-word footnote on energy terminology followed by a barely relevant detour into the history of particle accelerators. An illustration of the apparatus Coulomb used to derive his famous law is reproduced without any explanation of its operation.

A discussion of the evidence for Thomson's electron conjecture is interrupted by a 500-word digression concerning Pieter Zeeman and Hendrik Lorentz's attempts to measure the electron's mass-to-charge ratio. While they interrupt narrative flow, these passages can be skipped without endangering understanding. But Weinberg's narrative cannot be followed without an understanding the physics of motion, magnetism, electricity, and energy, the subjects of five "flashbacks," each of which seriously disrupts even the most determined reader.

The Final Theory

In *Dreams of a Final Theory* and the books that followed, Weinberg shows himself to be a man of culture. This breadth separates him from our other physicists. It is hard to imagine anyone else commenting that, after a string of theoretical and experimental successes, physicists "felt as did Siegfried after he tasted the dragon's blood, when he found to his surprise that he could understand the language of birds."[23] Or telling us what it is like to do theoretical physics: "I manipulate mathematical expressions and feel like Faust playing with his pentagrams before Mephistopheles arrives."[24] Or pointing out that Shakespeare trumps Einstein: "Often the director of a Shakespeare play chooses to leave out whole speeches. In the Olivier film version of *Hamlet*, Hamlet never says, 'Oh what a rogue and peasant slave am I! . . . 'And yet the performance works, because Shakespeare's plays are not spare, perfect structures like general relativity . . . ; they are big messy compositions whose messiness mirrors the complexity of life. That is part of the beauty of his plays, a beauty that to my taste is a higher order of beauty than the beauty of . . . general relativity."[25]

We can see his literary skills on full display in the second chapter of *Dreams of a Final Theory*, as step by step he climbs a ladder to the scientific sublime. The title of this tour de force, "On a Piece of Chalk," alludes to a famous lecture by Thomas Henry Huxley. Huxley's aim was to convince his audience that the chalk formations of England, most spectacularly evident in the white cliffs of Dover, took many millions of years to evolve, a span of time far exceeding the biblical six thousand years. Inspired by Huxley, Weinberg uses the same example to describe physicists' attempts to account for subatomic particles as the foundation for a "final theory," in our lexicon a "theory of everything."

A large segment of this chapter revolves around repeatedly asking why. Why is chalk white? Optics theory has the answer. The calcium carbonate of which chalk is composed strongly absorbs light in the spectrum that produces color. Why is that? Atomic theory provides the answer. The calcium carbonate molecules "do not happen to have any state that is particularly easy to jump to by absorbing photons of any color of visible light."[26] Why is that? Quantum mechanics has the answer: the probability of finding calcium carbonate atoms in certain quantum states. Why is that? The "standard model" of quantum mechanics has the answer. It explains the forces that arise as a result of the exchange of elementary particles, forces that yield the quantum states related to color production. At this point, Weinberg must stop drilling down; we are at the limit of what science can explain.

Weinberg then expands his discussion to encompass chemistry and biology. Chemical analysis shows that the calcium carbonate of chalk consists of calcium, carbon, and oxygen in the weight ratio 40:12:48. The atomic theory developed by chemists in the 19th century tells us why. It is because calcium carbonate is composed of one calcium, one carbon, and three oxygen atoms, that is, $CaCO_3$. Why that proportion? Atomic theory also has the answer. Since the weight of neutrons and protons is nearly equal and much, much greater than that of electrons, the weight of any given atom is roughly proportional to the number of its protons and neutrons. The atomic weights of 40 for calcium, 12 for carbon, and 16 for oxygen yield a weight ratio of 40:12:48 for $CaCO_3$. Why do the neutrons and protons form those atoms? The answer comes from the quantum properties of quarks, the primary constituents of protons and neutrons. Now Weinberg stops. We are again at the limit of what science can explain.

Finally, Weinberg turns to biology. In his lecture, Huxley noted that "chalk is composed of the skeletons of tiny animals who absorbed calcium salts and carbon dioxide from ancient seas and used these chemicals as raw materials to build little shells of calcium carbonate around their soft bodies."[27] Why the hard shell? Darwin and Wallace have the answer: to survive in a sea full of predators. Why do these animals possess that feature? The answer is natural selection. Why do organisms inherit these favorable traits? The answer comes from molecular biology: the double-helix DNA molecule stores and transmits genetic information. By continuing to drill down, we return again to the limit of what science can explain.

The final theory that undergirds the explanations of biology and chemistry, that explains these explanations, is, as Weinberg freely admits, not yet a reality. There are serious limitations to realizing that dream: most notably, the enormous complexity involved in transforming these diverse theories into a set of fundamental equations. Physicists are nowhere near using the equations of quantum mechanics or the Standard Model to explain the workings of DNA, much less human consciousness or behavior. More broadly, as pointed out by Roger Penrose, physicists are a long way from explaining "how small-scale quantum processes can add up, for large and complicated systems, to the almost classical behavior of macroscopic bodies."[28] Weinberg wisely counsels that "the final theory may be centuries away and may turn out to be totally different from anything we can now imagine."[29] Yet he also makes clear that physicists appear to be within striking distance of a final theory in a limited sense, a theory uniting the four forces of nature—the strong, weak, electromagnetic, and gravity. This theory would be a quantum jump beyond the Standard Model.

Defending Science

Some scientists have viewed the claims of radical sociologists and philosophers of science—claims that science is largely or entirely a social construction—as attempts to diminish, tarnish, or distort their disciplines, to turn the scientific sublime into a collective hallucination. In 1996, one physicist, Alan Sokol, struck back. He published an article in *Social Text* that its editors accepted only because they did not recognize a lampoon for what it was. When Sokol revealed himself, the science wars began. Weinberg soon entered the fray. While science is a manifestation of culture, its findings, he insisted, were beyond culture; they tracked enduring truths. His defense involved an attack on the work of his colleague and friend Thomas Kuhn, a disagreement difficult to fathom. Kuhn and Weinberg both held doctorates in physics. Like Weinberg, Kuhn believed in scientific progress; like Weinberg, he believed that, despite changes as radical as the shift from the geocentric universe of Ptolemy to the heliocentric universe of Copernicus, some good science remained good; like Weinberg, he thought the deep-seated relativism of some prominent sociologists and philosophers was sheer nonsense.

There is one point of difference that accounts for Weinberg's attack. Kuhn confesses that he has difficulty conceiving of science as closing in on the truth; he feels that we "may have to relinquish the notion" that we are moving closer and closer to the way the world is.[30] Kuhn can't make sense of the expression "closer to truth"; it does seem distinctly odd to say that Einstein is closer to the truth than Newton when we cannot say how far away Newton was or how much nearer Einstein is. According to Weinberg, however, closing in on truth is what the sciences do. He is certain because "if scientists are talking about something real, then what they say is either true or false."[31]

So is the Standard Model of quantum mechanics true or false? If the Large Hadron Collider had revealed that there was no Higgs particle, the Standard Model would have been proven false. Still, it would have continued to make accurate predictions, the very criterion Weinberg espouses when he says that "a theory is taken as a success if it is based on simple general principles and does a good job of accounting for experimental data in a natural way."[32] Indeed, Weinberg is convinced that the Standard Model must be measured by just this criterion. It is, he says, "an 'effective field theory,' a low energy approximation of some unknown fundamental theory that may not involve fields at all."[33] This view of science imagines its truth not as a destination but as an endless road, paved with probable conjectures, a sublime that continues to elude our grasp. It says that while it is nonsense to assert that we are getting closer to a horizon, it is not nonsense to say that as we travel forward, the territory of our understanding continually expands. This is Kuhn's view. Were Weinberg thinking clearly, I think that it would be his.

There is another problem with Weinberg's defense of science, his insistence that the conjectures of elementary particle physics are "more fundamental" than those of any other science, that his sublime is more sublime than their sublime. At the end of the day, he feels, all sciences will be his science, quantum mechanics.[34] This is an imperialism that puts him in direct conflict with another Nobel laureate, Philip Anderson, a condensed matter physicist. Anderson thinks that high-temperature superconductivity— the enhanced conductivity at temperatures that are still cold but well above absolute zero—is a consequence of broken symmetry, a concept he borrows from quantum mechanics. He freely acknowledges the debt. But to him, this broken symmetry means that high-temperature superconductivity cannot be deduced from the equations of quantum mechanics. Although Weinberg cannot but agree, he feels nevertheless that "when condensed matter physicists finally explain high temperature superconductivity— whatever brilliant ideas have to be invented along the way—in the end the explanation will take the form of a mathematical demonstration that deduces the existence of this phenomenon from *known* properties of electrons, photons, and atomic nuclei."[35] Anderson begs to differ. It is simply not true that complex states can invariably be deduced from simple ones:

> The state of a really big system does not at all have to have the symmetry of the laws which govern it; in fact, it usually has less symmetry. The outstanding example of this is the crystal: Built from a substrate of atoms and space according to laws which express the perfect homogeneity of space,

the crystal suddenly and unpredictably displays an entirely new and very beautiful symmetry.[36]

The implications of this quarrel are not merely technical: the correct view of the route to the sublime has practical consequences. Anderson alleges a long-standing campaign of denigration. He asserts that T. D. Lee and Samuel C. C. Ting, both Nobel laureates, made sure that Columbia University hired no condensed matter physicists.[37]

Conclusion

Steven Weinberg is a master at communicating the conjectural sublime, summoning up the early universe, in which subatomic particles turn into stubborn facts. His first book, *The First Three Minutes*, tells of a time when rules governing the micro-world were everywhere in force, a conjectural past founded on particle physics. Next, in *The Discovery of Subatomic Particles*, he tells us about J. J. Thomson's discovery of the electron, an experiment that turned a conjectural into a real particle, the beginning of a new era in physics. In *Dreams of a Final Theory*, his third popular science book, he argues for the continuation of Thomson's program of discovery, in which bold conjectures lead inevitably to a final theory, the fundamental principles according to which the cosmos came into existence. The next best step in realizing this dream of a final theory, he strongly avers, is the building of the world's most expensive machine—the Superconducting Supercollider, a particle accelerator designed to turn the conjectures of a final theory into the facts of science, a dream transformed into reality in the Large Hadron Collider, a subject that Lisa Randall investigates. Weinberg's latest, *To Explain the World*, traces the transformations of the scientific sublime over thousand years. His is a story that unfortunately slights the contributions of such justly famous contributors to science as Francis Bacon and René Descartes, men who, as it turns out, were in some respects mistaken, as mistaken as Steven Weinberg may turn out to be decades from now. While he may eventually cease to be correct, he will not cease to be a great physicist, a man who reached out to a sublime that may—just—have eluded him.

CHAPTER 4 | Lisa Randall
The Technological Sublime

Singing my days,
Singing the great achievements of the present,
Singing the strong light works of engineers,
Our modern wonders, (the antique ponderous Seven outvied).

—WALT WHITMAN

IN 2008 A rap video by Kate McAlpine went viral (nearly eight million views at present). Not your typical rap video, it takes place in the tunnel of the Large Hadron Collider and on the grounds 100 feet above. During the performance, the computer-generated voice of Stephen Hawking chimes in as part of a periodic call and response. Throughout, the lyrics are replete with technical terms like "protons," "lead ions," "antimatter," "black holes," "dark matter," "Higgs boson," "Standard Model," "graviton," "top quark," and acronyms like "ALICE," "ATLAS," and "CMS." Here is the central refrain:

The LHC accelerates the protons and the lead
And the things that it discovers will rock you in the head.
The Higgs boson, that's the one that everybody talks about
And it's the one sure thing that this machine will sort out.[1]

McAlpine's was a prophesy that proved right on target. In 2016, François Englert and Peter Higgs won the Nobel Prize in physics for a conjecture they had made over a half century earlier, a mathematically driven leap of faith that became a scientific fact when the Higgs boson was detected—a hitherto mysterious but absolutely central member of the particle zoo. It was a discovery that confirmed the otherwise highly confirmed Standard Model, the explanatory centerpiece of the quantum world. At five billion dollars, the detector of the Higgs, the Large Hadron Collider, is the most expensive scientific apparatus ever built. It is a Mount Everest of machines, the apotheosis of the technological sublime. This form of sublimity is near the center of Lisa Randall's professional life, the only means by which her deepest conjectures about the universe can be demonstrated. Hers is a flight into the scientific stratosphere tethered to events that she hopes will be observed by two incarnations of the technological sublime: the Large Hadron Collider or the GAIA satellite.

The Higgs Boson

When the UK funding for the Large Hadron Collider was still in question, Science Minister William Waldegrave challenged British physicists, telling them "that if anyone could explain what all the fuss was about, in plain English, on one sheet of paper, then he would reward that person with a bottle of vintage champagne."[2] Had Lisa Randall been a contestant, she might well have partaken of the Veuve Clicquot, for she brings down to earth scientific speculation at its most abstruse. Dinner table settings are her analogy for symmetry, a central characteristic of the micro-world. Dinner table symmetry is perfect until someone decides to reach for the glass:

> At that moment the left-right symmetry is spontaneously broken. No law of physics dictates that anyone has to choose left or right. But one has to be chosen; after that, left and right are no longer the same in that there is no longer a symmetry that interchanges the two.[3]

Quantum mechanical symmetries guarantee "that theoretical predictions . . . agree with experiments";[4] unfortunately, they also guarantee that they disagree. They predict nonsense: they tell us that some elementary particles—those we know have mass—are massless. For these particles to exist, the symmetry of the Standard Model must be broken, the task of the Higgs field, for which the discovery of the Higgs particle would be incontrovertible evidence. Randall compares the Higgs field, which permeates "empty" space, to a "viscous fluid":

> Particles that carry this charge, such as the weak gauge bosons and Standard Model quarks and leptons, can interact with this fluid, and these interactions

slow them down. This slowing down then corresponds to the particles acquiring mass.[5]

The discovery of the Higgs and its associated field is a triumph of the Large Hadron Collider, an embodiment of the technological sublime. According to historian David Nye, we experience this sublime when we meet a startling man-made object like the Hoover Dam, an encounter that "disrupts ordinary perception and astonishes the senses, forcing the observer to grapple mentally with its immensity and power."[6] Nye's examples are drawn from American technology: the transcontinental railroad, Erie Canal, atomic bomb, Apollo spacecraft, and Brooklyn Bridge. Of the Brooklyn Bridge, Nye notes the many artistic renderings of its "monumental stone piers and ethereal spun-steel cables" and quotes one of the most famous painters of this sublime architecture, Joseph Stella: "Many nights . . . I stood on the bridge. . . . I felt deeply moved, as if on the threshold of a new religion or in the presence of a new DIVINITY."[7] Poet Hart Crane expressed similar lofty sentiments in his homage:

O Sleepless as the river under thee,
Vaulting the sea, the prairies' dreaming sod,
Unto us lowliest sometime sweep, descend
And of the curveship lend a myth to God.[8]

Nye maintains that the American technological sublime arose in the wave of 19th-century boasting: America had "the *biggest* waterfall, the *longest* railway bridge, or the *grandest* canyon."[9] But boasting was not just boasting:

The sublime taps into fundamental hopes and fears. It is not a social residue, created by economic and political forces, though both can inflect its meaning. Rather, it is an essentially religious feeling, aroused by the confrontation with impressive objects, such as Niagara Falls, the Grand Canyon, the New York skyline, the Golden Gate Bridge, or the earth-shaking launch of a space shuttle. The technological sublime is an integral part of contemporary consciousness, and its emergence and exfoliation into several distinct forms during the past [three] centuries is inscribed within public life. In a physical world that is increasingly desacralized, the sublime represents a way to reinvest the landscape and the works of men with transcendental significance.[10]

To describe the Large Hadron Collider, Randall employs the vocabulary of the sublime. Of her first visit she says, "I was surprised at the sense of awe it inspired—this in spite of my having visited particle colliders and detectors many times before. Its scale was simply different. . . . The complexity, coherence, and magnitude, as well as the crisscrossing lines and colors, are hard to convey in words. The impression is awe inspiring."[11] The LHC is "the *most important* experimental machine for particle physicists"[12] with a long list of arresting properties. The first is the triumph over time. The collider will be able to simulate events that occurred in the "first trillionth of a millisecond after the Big Bang."[13] The second is the triumph over space, the investigation of the tiniest components of the universe: "Incredibly small sizes—on the order of a tenth of a thousandth of a trillionth of a millimeter . . . a factor of ten smaller in size than anything any experiment has ever looked at before."[14] Other superlatives concern energy: "up to seven times the energy of the highest existing collider"; temperature: "even colder than outer space"; magnetic field: "100,000 times stronger than the Earth's"; and cost: "the most expensive machine ever built" (five billion dollars, with operating costs of one billion a year). At $385 million in today's dollars, the Brooklyn Bridge was a bargain.

The story of the Higgs discovery is told in the documentary film *Particle Fever*, the joint effort of director Mark Levinson, who holds a PhD in physics, and Walter Murch, the winner of two Oscars as a film editor. A masterpiece of editing, *Particle Fever* begins with cows munching grass in a Swiss meadow just outside Geneva. In the near distance, in the midst of this bucolic tranquility, we see CERN, the Conseil Européen pour la Recherche Nucléaire, the location of the Large Hadron Collider, a machine in a garden. We see experimental physics postdoc Monica Dunford on her way to the collider, cycling past Route Marie Curie and Route R. Feynman, an instant starlet who characterizes the collider as "a five-story Swiss watch," one that will, when the Higgs experiment is carried out, create a child's sense of excitement at a forthcoming birthday party where "there's goin' to be cake."

Theoretical physicist David Kaplan lets us in on the high stakes involved in this experiment. Theoretical physics is "at a fork in the road" in which "an entire field is hanging on a single event," on which 10,000 people are working in concert, a remarkable effort in which Iranian and Israeli physicists labor side by side. When asked about the point of these extraordinary efforts, Kaplan has two answers. To the general public and the media, he says that they are in the business of discovering the origin of the universe. The real answer is shared only among physicists: the revelation of the basic laws of nature. Asked by an economist to place a value on this enterprise, Kaplan says, "It could be nothing"—nothing "but understanding everything."

That the experiment could really be nothing—that it could fail—haunts the staff, who would have preferred to conduct it between midnight and dawn. In fact, the Higgs was discovered in broad daylight in the full glare of publicity, a scary prospect. In the event, certainty was achieved within 5 sigma, a one-in-35-million chance of error. During the announcement ceremony, the technological, theoretical, and moral sublime merged. Peter Higgs rose from his seat to give thanks that he has lived long enough to see his dream come true.

It was the dream of an unlikely Nobel laureate. Higgs had trouble getting his papers in print. No one believed him; no one was interested in what he had to say. When he was scheduled to speak at Harvard, "members of the audience were looking forward to tearing apart this idiot who thought he could get around the Goldstone theorem [showing that his result was impossible]."[15] Higgs was even in danger of losing his job because a lack of productivity: "I was an embarrassment to the department when they did research assessment exercises. A message would go round the department: 'Please give a list of your recent publications.' And I would send back a statement: 'None.'"[16] Clearly, we have an administration focused on quantity, not on quality; worse, we have theoretical physicists closed to new ideas and deaf to the truth.

What, exactly, had happened in the Collider on July 4, 2012, the date of the discovery? The Higgs experiment began when proton beams clashed, creating a billion events a second. To help us understand the meaning of this encounter, Randall relies on analogy:

> The collisions of protons are analogous in some respects to two beanbags colliding. Because beanbags are soft, most of the time they wilt and hang and don't do anything interesting during the collision. But occasionally when beanbags bang together, individual beans hit each other with great force—maybe even so much so that individual beans collide and the bags themselves break. In that case, individual colliding beans will fly off dramatically, since they are hard and collide with more localized energy, while the rest of the beans will fly along in the direction in which they started.[17]

Of course, protons are not beanbags. Upon collision, they convert into other forms of matter or into energy. It is from these decay products that the existence of the Higgs boson, the signature of the Higgs field, is inferred. The result turns out to be ambiguous; it points in no definite theoretical direction. While the discovery supports the Standard Model, the new particle is too light to point definitively in the direction of a multiverse, Brian Greene's theory in which our universe is one of many, and too heavy to point in the direction of supersymmetry, an upgrade of the Standard Model. Still, there is little doubt that this discovery is the capstone of the "extraordinary era in which we live. It is altogether new. The world has seen nothing like it before. I will not pretend, no one can pretend, to discern the end; but everybody knows that the age is remarkable for scientific research into the heavens, the earth, and what it beneath the earth. . . . The ancients saw nothing like it. The moderns have seen nothing like it till the present generation."[18] It is not Peter Higgs speaking, or Monica Dunford or David Kaplan or Lisa Randall. It is Daniel Webster; the date is August 28, 1847, the day the Northern Railroad opened.

Revising the Standard Model: Supersymmetry

Higgs had argued successfully that his boson and its associated field were necessary to maintain the mathematical consistency of the reigning theory of quantum physics, the Standard Model. Nobel physicist Frank Wilczek describes this model with superlatives typical of the sublime: "one of the grandest achievements—I would argue, *the* grandest achievement—of human thought and striving."[19] Randall defines this model with admirable clarity; it describes "the most elementary components that are known to matter, such as quarks, leptons (like the electron), and the three nongravitational forces through which they interact—electromagnetism, the weak nuclear force, and the strong nuclear force. . . . [It] gives completely consistent predictions of all known particle phenomena at the level of precision of a fraction of a percent."[20]

Randall visualizes the relationship among the three forces and their associated fundamental particles by means of a tabular arrangement reminiscent of the periodic table, an ingenious representation of a model that is, in the words Adam Smith applied to Newton's theory of gravity, "an immense chain of the most important and sublime truths, all closely connected together."[21] While this table is a triumph of design, the grandeur of the vision it encapsulates is revealed only with dogged effort. Although Randall avoids equations, she cannot avoid inherent complications that require careful study. The many required traversals from table to text required for comprehension may well exceed the limits of tolerance of most of her readers.

At the top of figure 4.1 are two types of fundamental particles, *quarks* and *leptons*. The row at the bottom displays the *gauge bosons*, fundamental

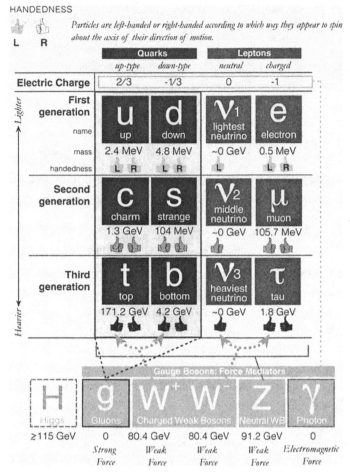

HANDEDNESS

L   R   *Particles are left-handed or right-handed according to which way they appear to spin about the axis of their direction of motion.*

	Quarks		Leptons	
	up-type	down-type	neutral	charged
Electric Charge	2/3	-1/3	0	-1
First generation name	**u** up	**d** down	**ν₁** lightest neutrino	**e** electron
mass	2.4 MeV	4.8 MeV	~0 GeV	0.5 MeV
handedness	L R	L R	L	L R
Second generation	**c** charm	**s** strange	**ν₂** middle neutrino	**μ** muon
	1.3 GeV	104 MeV	~0 GeV	105.7 MeV
Third generation	**t** top	**b** bottom	**ν₃** heaviest neutrino	**τ** tau
	171.2 GeV	4.2 GeV	~0 GeV	1.8 GeV

Gauge Bosons: Force Mediators

H	g	W⁺ W⁻	Z	γ
Higgs	Gluons	Charged Weak Bosons	Neutral WB	Photon
≥115 GeV	0	80.4 GeV 80.4 GeV	91.2 GeV	0
	Strong Force	*Weak Force Weak Force*	*Weak Force*	*Electromagnetic Force*

FIGURE 4.1 Representation of Standard Model particles and interactions.

SOURCE: Lisa Randall, *Knocking on Heaven's Door: How Physics and Scientific Thinking Illuminate the Universe and the Modern World* (New York: Ecco, 2012), 242.

particles that interact with the quarks and leptons to yield the three non-gravitational forces—strong, weak, and electromagnetic. Gluons communicate the strong force; W⁺, W⁻, and Z particles, the weak force; the photon, electromagnetic force. Various lines in the figure—dotted, dashed, and solid—link the quarks and leptons to their respective force mediators. For the two sets of six quarks and leptons, the figure also provides the name and key quantum properties that differentiate them: electric charge, spin, and mass.

Randall defines each of the particles. In the first row "the proton contains two up quarks and one down quark, while the neutron contains two down quarks and one up quark."[22] The up and down quarks, together with the electron, constitute "the fundamental building blocks" of atoms.[23] Moving to next two rows, she explains that

> heavier quarks called *charm* and *strange* and *top* and *bottom* exist as well. Their masses are given in electron volts, a standard practice. As with the charged and neutral leptons, the heavier quarks have charges identical to their lighter counterparts [either 2/3 or −1/3, as shown in bar at top]—the up and down quarks.[24]

The six flavors of quark experience all three non-gravitational forces, as indicated by the various lines linking these quarks with the gluons, W^+, W^-, Z, and the photon.

Randall defines leptons as "particles such as the electron that don't experience the strong force."[25] For that reason, unlike quarks, they do not get bound into protons and neutrons. In the same column as the electron are the *muon* and *tau* particles, "two heavier versions of the electron, which have the same charge [indicated by −1 in bar at top] but much bigger masses." We see this when we read the mass values given in each cell of the far-right column. The other type of lepton, the neutrinos, "don't carry a charge at all [signified by 0 in bar in top bar] . . . [and] are extremely tiny but nonvanishing masses [~0 GeV]" that experience only the weak force. Randall defines each gauge boson in the bottom row simply as "a fundamental, elementary particle that is responsible for communicating a particular force.[26] Randall refers to these as "force mediators."

This information-packed figure also displays the "handedness" of the quarks and leptons, that is, the direction of their spin, represented by a left- or right-hand icon. A comparison of the spin for all the leptons and quarks immediately identifies an important asymmetry: the weak force acts only on particles with the left-handed icons, as indicated by the arrows with dashed lines anchored in the force mediator row. The spin of the three neutrinos is only to the left.

All that remains to explain is the lonely Higgs particle at the bottom left. It is "not just a new particle, but a new type of particle."[27] The associated Higgs field "is something like a charge." It permeates the entire universe and gives mass to leptons, quarks, and other bosons. The Higgs is central to the story of the origin of matter in the very early universe, creating a phase transition:

> The idea—whatever we call it—is that a phase transition (perhaps like the phase transition of liquid water bubbling into gaseous steam) took place

that actually changed the nature of the universe. Whereas early on, particles had no mass and zipped around at the speed of light, later on—after this phase transition involving the so-called Higgs field—particles had mass and traveled more slowly. The Higgs mechanism tells how elementary particles go from having zero mass in the absence of the Higgs field to the nonzero masses we have measured in experiments.[28]

To have mastered this arcane terminology is to have penetrated to a level deeper than the periodic table—to have discovered the elements of the elements. But while the periodic table is organized in a way any high school student can easily grasp, the table in figure 4.1 is not. On close inspection, a critic might well exclaim: The Higgs boson doesn't fit, stuck as it is in a convenient corner. The critic might also ask: Why all the different masses; is there some pattern or organizing principle behind them analogous to periodicity? Moreover, the complicated relationships among the leptons and quarks and bosons do not appear to be part of any grand design that favors consistency in symmetry and pattern. As Wilczek, a major contributor to the Standard Model, admits, "It's a kludge, for sure, and a harsh critic might call it a mess."[29]

One answer to that criticism is a transformation of figure 4.1. In figure 4.2, the Higgs moves into a row of its own with three adjacent companions. Moreover, there is, for "every fundamental particle of the Standard Model—electrons, quarks, and so on— . . . , a partner in the form of a particle with similar interactions but different quantum mechanical properties."[30] In this new "supersymmetric" arrangement, each of the quarks and leptons is paired with a gauge boson, a force mediator. The new families add "s" to the Standard Model name: the substance particle *up quark*—a basic component of protons and neutrons—becomes the force mediator *sup squark*. The once isolated Higgs boson now has a family with supersymmetric partners. Figure 4.1 is a representation of what is known; figure 4.2, what might be. The contrast between figure 4.1, where the Higgs stands out like a sore thumb, and figure 4.2, where it does not, is comprehended in a flash, a visual exclamation point.

Still no more than speculation, supersymmetry improves on the Standard Model. In Brian Greene's words, "It turns out that the consistency of the *supersymmetric standard model*—the standard model augmented by all of the superpartner particles—no longer relies on the uncomfortably delicate numerical adjustments of the ordinary standard model."[31] Supersymmetry also predicts a host of new Higgs bosons and supersymmetric particles, which if they exist, should be detectable in the Large Hadron Collider.

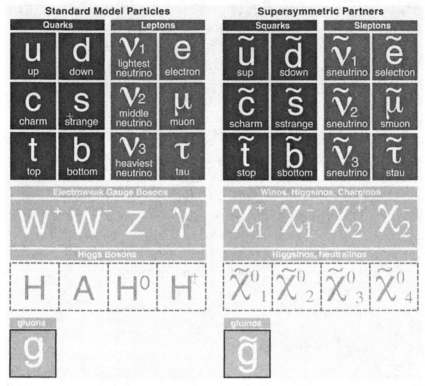

FIGURE 4.2 Transformation of the Standard Model with supersymmetric partners and new Higgs bosons.

SOURCE: Lisa Randall, *Knocking on Heaven's Door: How Physics and Scientific Thinking Illuminate the Universe and the Modern World* (New York: Ecco, 2012), 303.

Of this new theory, Randall says, "Since the 1970s, many physicists have considered the existence of supersymmetric theories so beautiful and surprising that they believe that [they have] to exist in nature."[32]

Solving the Hierarchy Problem

When e. e. cummings told us that "there's a hell of a good universe next door; let's go,"[33] he certainly did not have in mind Randall's alternative to the supersymmetry model, a theory that posits a fifth dimension rather than a gaggle of new particles. To communicate her theory's abstruse mathematics, to bring the scientific sublime down to earth, Randall begins by describing the physics about which there is consensus: general relativity, quantum field theory, the Standard Model. The first explains the force of gravity; the second, the weak and strong forces; the third, "all known particles and non-gravitational forces and the interactions among them."[34] Her introduction leads to the hierarchy problem her theory is designed to solve: Why is the force of gravity so many orders of magnitude weaker than the three? To dramatize the difference, Randall has recourse to a quantitative analogy:

> This gravitational attraction [between two electrons] is about a hundred million trillion trillion trillion times weaker than the electric repulsion between the electron pair. . . . That's an enormous number—it's comparable to the number of times you could lay the island of Manhattan end to end in the extent of the observable universe.[35]

Devised with co-author Raman Sundrum, Randall's solution requires one additional dimension—four dimensions of space and one of time. It predicts the existence of a particle, the Kaluza-Klein, which is yet to be detected. The solution is contained in two articles published in *Physical Review Letters*.[36] Here is the abstract to the first, the technical terms in bold face:

> We propose a new **higher-dimensional mechanism** for solving the **hierarchy problem**. The **weak scale** is generated from the **Planck scale** through an **exponential hierarchy**. However, this exponential arises not from **gauge interactions** but from the **background metric** (which is a slice of **AdS5 spacetime**). We demonstrate a simple explicit example of this mechanism with two **3-branes**, one of which contains the **Standard Model fields**. The phenomenology of these models is new and dramatic.[37]

Warped Passages translates this rebarbative prose into a set of ingenious images that shed light on key technical terms, making concrete concepts whose mathematics general readers could not possibly follow.

Figure 4.3 visualizes the two words ending the abstract's first sentence: "We propose a new higher-dimensional mechanism for solving the **hierarchy problem**." The vertical double arrow points to the enormous gap in mass, expressed as billions of electron volts (GeV). The associated length scale is in centimeters. Gravity (Planck scale mass) is 10,000,000,000,000,000 times weaker than the other three forces (weak scale mass, proton mass, and electron mass).

Randall conveys this degree of difference by means of everyday analogies: "This ratio is greater than the number of minutes that have passed since the big bang; it's about a thousand times the number of marbles you

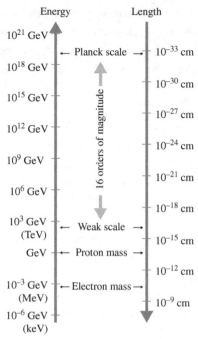

FIGURE 4.3 Hierarchy problem.

SOURCE: Lisa Randall, *Warped Passages: Unravelling the Mysteries of the Universe's Hidden Dimensions* (New York: Harper Perennial, 2005), figs. 63 and 80 (pp. 255, 392). Copyright © by Lisa Randall. Reprinted by permission of HarperCollins Publishers.

can line up from the Earth to the Sun."[38] For those still perplexed by the hierarchy problem, Randall offers an analogy involving twin brothers:

> Imagine that you visit a very tall friend, and discover that although he is 6'5" tall, his fraternal twin is only 4'11". That would be surprising. . . . Now imagine something ever more bizarre: you walk into your friend's house and find that your friend's brother is ten times smaller or ten times bigger. . . . Just as you have good reason to expect similar heights among family members, particle physicists have valid scientific reasons to expect similar masses among particles in a single theory, such as the Grand Unified Theory. But in a GUT the masses are not all the same: even those particles that experience similar forces have enormously different masses. And not by a factor of ten: the discrepancy between masses is more like a factor of ten trillion.[39]

This vast difference in magnitude between gravity and the other forces has posed so far insurmountable problems for physicists who would like to generate a set of equations that unites all four forces into a single theory.

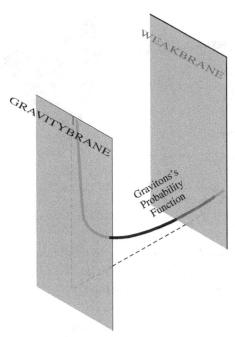

FIGURE 4.4 Exponential hierarchy.

SOURCE: Lisa Randall, *Warped Passages: Unravelling the Mysteries of the Universe's Hidden Dimensions* (New York: Harper Perennial, 2005), 392. Copyright © by Lisa Randall. Reprinted by permission of HarperCollins Publishers.

Randall and Sundrum's abstract asserts that they have found a "higher dimensional mechanism" that solves the hierarchy problem: "The weak scale is generated from the Planck scale through an **exponential hierarchy**." Figure 4.4 imagines this hierarchy, one that entails a hidden fifth dimension adjacent to our four. We see two parallel universes, or "branes," described by *New York Times* reporter Dennis Overbye as islands "of three dimensions floating in a sea of higher dimension, like a bubble in a sea."[40] One is the "Weakbrane," home to *Homo sapiens* and all the known fundamental particles. The other is the "Gravitybrane," where the force of gravity is superstrong. These branes are separated by a fifth dimension in which the force of gravity decreases exponentially: gravity is some sixteen orders of magnitude stronger on or near the Gravitybrane. In the words of the abstract, "The weak scale is generated from the Planck scale through an exponential hierarchy."

The abstract's third sentence elaborates on the authors' solution: "However, this exponential arises not from gauge interactions but from the background metric (which is a slice of **AdS5 spacetime**)." What is AdS5? Figure 4.5 shows that within the fifth dimension, spacetime is warped. Physicists call this tunnel-like image an "anti de Sitter

FIGURE 4.5 Warped AdS5 space-time.

five-dimensional space," or "AdS5." The image on the left represents the "warped passage" in the title of Randall's first book. She explains the image on the right by means of an analogy:

> We could slice the funnel into flat sheets with a cleaver, but the surface of the funnel is clearly curved. This is similar to the curved five-dimensional spacetime we're considering. But the analogy isn't perfect, because the boundary of the funnel, the funnel's surface, is the only space where it's curved, whereas in the warped spacetime the curvature is everywhere.[41]

The abstract's fourth sentence elaborates the authors' solution: "We demonstrate a simple explicit example of this mechanism with two **3-branes**, one of which contains the **Standard Model fields.**" Figure 4.6, an elaboration of figure 4.4, visualizes the terms in bold. The "two 3-branes" are the Weakbrane and Gravitybrane, the number three signifying the three spatial dimensions. In the "bulk," the fourth spatial dimension, the force of gravity changes exponentially between the two branes. Stuck on the Weakbrane, we see the Standard Model particles—W and Z bosons that mediate the weak force (symbolized by the sun), gluons that mediate the strong force (the atom icon), and photons that mediate the electromagnetic force (the magnet icon).

Figures 4.4–4.6 represent a solution to the hierarchy problem visualized in figure 4.3, a triumph in translating a concept only a few theoretical physicists understand into a form any intelligent person can grasp. The whole of the authors' abstract builds to the bold claim of their last sentence: "The phenomenology of these models is new and dramatic." It is a drama Randall realizes not only in her co-authored paper and in *Warped Passages* but in a medium unique among the works of science

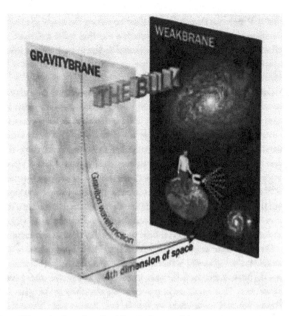

FIGURE 4.6 Possible solution for the hierarchy problem.

SOURCE: Lisa Randall, *Warped Passages: Unravelling the Mysteries of the Universe's Hidden Dimensions* (New York: Harper Perennial, 2005), 388. Copyright © by Lisa Randall. Reprinted by permission of HarperCollins Publishers.

popularizers, the libretto of Hèctor Parra's two-person opera *Hypermusic Prologue: A Projective Opera in Seven Planes.*[42]

Dark Matter

In *Dark Matter and the Dinosaurs: The Astounding Interconnectedness of the Universe,* Lisa Randall addresses another puzzle that tantalizes physicists. While invisible, dark matter "constitutes 85 percent of the matter in the Universe."[43] Although wholly transparent, we know of its existence because of "the expansion of the Universe, the path of light rays passing to us from distance objects, the orbits of stars around the centers of galaxies, and many other measurable phenomena."[44] While confident of its existence, physicists "don't yet know what it is" and have yet to detect it.[45] Randall employs an analogy to explain why she believes in what she has never observed:

> When walking down the sidewalks of Manhattan or driving along the streets of Hollywood, you sometimes sense that a famous person is near. Even if you don't see George Clooney directly, the disruptive traffic generated by the waiting crowd armed with cell phones and cameras suffices to alert you

to a celebrity's proximity. . . .We don't see dark matter, but—like the celebrity . . . —it influences its surroundings.[46]

How does dark matter account for the disappearance of the dinosaurs, the puzzle Randall is determined to solve? There is one thing we know about dark matter: it surrounds the Milky Way, a "large spherical halo—about 650,000 light-years wide."[47] Building on this consensual science, Randall creates a speculative scenario in which "a small component of dark matter . . . interacts through nongravitational forces,"[48] an event that generates a vast dark-matter disk embedded within this halo: "Just as ordinary matter cools and collapses, so would the interacting component of dark matter. Moreover, because of conserved angular momentum, which prevents collapse in all but the vertical direction, the interacting dark matter would collapse into a disk."[49]

In Randall's dinosaur extinction scenario, sixty-six million years ago in one of its periodic oscillations, the solar system passed near this dark matter disk, dislodging a slew of long-period comets from the far-distant Oort Cloud.[50] One of these, "three times the width of Manhattan" and "bigger than the height of Mount Everest, crashed into the Yukatán Peninsula," its velocity "500 times faster than a vehicle on an autobahn." Most impressive was its destructive force, "equivalent of up to 100 trillion tons of TNT, more than a billion times greater than the atom bombs that destroyed Hiroshima and Nagasaki." Its effects are vividly described:

> Extreme winds and waves raged, and huge tsunamis radiated. . . . Tidal waves also would have appeared on the opposite side of the globe, triggered by perhaps the most massive earthquake the Earth has ever experienced. . . . Trillion of tons of material would have been ejected. . . . Fires would have raged everywhere and Earth's surface would have literally been cooked. . . . More than half of the world's biomass was incinerated within months. . . . The oceans didn't recover for hundreds of thousands of years and most likely saw destructive influences for at least a half million to a million years afterward. . . . It seems that no living creature survived that was heavier than about 25 kilograms—about the weight of a midsize dog.[51]

According to Steven Weinberg, "any explanation of the present forms of life on earth must take into account the extinction of the dinosaurs . . . currently explained by the impact of a comet, but no one will ever be able to explain why a comet happened to hit the earth at just that time."[52] Although Randall has set out to prove Weinberg wrong, she reminds us that her solution is based on two conjectures: highly "speculative ideas about dark matter"[53] and the conjectured periodic release of comets from the Oort Cloud. Yet hers is not mere speculation. She and her colleagues have published three articles on the formation of the hypothesized dark matter disk and its triggering of periodic comet impacts.[54] While "statistical evidence is not overwhelming,"[55] at least one aspect of their hypothesis will be put to the test by the GAIA satellite, which is, like the Large Hadron Collider, a realization of the technological sublime. GAIA is designed to "measure the positions and velocities of a billion stars of the Milky Way,"[56] measurements that will tell whether those stars are moving in a way consistent with a dark matter disk.

Conclusion

Throughout Randall's books, we encounter two successful communicative strategies: images that make the invisible visible and analogies that make abstract concepts concrete. It is these strategies that give Randall's readers the Ariadne's thread with which to navigate the maze of current theoretical physics, a speculative sublime that depends for validation on such technological marvels as the Large Hadron Collider and the GAIA satellite. By this skillful communicative exploitation, Randall has brought to the art of popular writing a new level of clarity for which many readers are grateful—but not all. One Amazon purchaser laments: "I think Lisa Randall is a very, very smart woman who knows her subject who, like most brilliant minds, is awful in relaying what she knows to others, whether it be by verbal or written word. Honestly, this book is painful to read." For such readers, while her vision of technology and physics may be sublime, the experience of her prose is not. At least in the case of her exposition of the Standard Model, she seems to have reached the limit of what is possible without recourse to the language of mathematics.

CHAPTER 5

Brian Greene

The Speculative Sublime

Why, sometimes I've believed as many as six impossible things before breakfast.

—THE WHITE QUEEN TO ALICE, in Lewis Carroll, *Through the Looking-Glass*

CHARLES DODGSON WARNED a child correspondent of the dangers of living in the looking-glass world of mathematicians like himself, the high price of consistently believing "six impossible things before breakfast":

> Don't be in such a hurry to believe next time—I'll tell you why—If you set to work to believe everything you will tire out the muscles of the mind, and then you'll be so weak you won't be able to believe the simplest true things. Only last week a friend of mine set to work to believe Jack-the-giant-killer. He managed to do it, but he was so exhausted by it that when I told him it was raining (which was true) he *couldn't* believe it, but rushed out into the street without his umbrella, the consequence of which was his hair got seriously damp, and one curl didn't recover its right shape for nearly two days.[1]

In all his books, Brian Greene is our tour guide on a journey into his particular looking-glass world—string theory, an exercise in the speculative sublime, a sublime only for aficionados, certainly not for you and me. Here is the abstract of an article cited a respectable 201 times:

> We show that a string-inspired Planck scale modification of general relativity can have observable cosmological effects. Specifically, we present a complete analysis of the inflationary perturbation spectrum produced by

a phenomenological Lagrangian that has a standard form on large scales but incorporates a string-inspired short distance cutoff, and find a deviation from the standard result. We use the de Sitter calculation as the basis of a qualitative analysis of other inflationary backgrounds, arguing that in these cases the cutoff could have a more pronounced effect, changing the shape of the spectrum. Moreover, the computational approach developed here can be used to provide unambiguous calculations of the perturbation spectrum in other heuristic models that modify trans-Planckian physics and thereby determine their impact on the inflationary perturbation spectrum. Finally, we argue that this model may provide an exception to constraints, recently proposed by Tanaka and Starobinsky, on the ability of Planck-scale physics to modify the cosmological spectrum.[2]

If this is not clear—it might as well be written in Hittite—we can always refer to Greene's mathematics:

The parameter β is related to the minimum distance Δx_{min} by $\Delta x_{min} \sim \sqrt{\beta}$ A Lagrangian suggested by Eq. (1) was discussed in Ref. [5]. In this model, the tensor mode $u_{\tilde{k}}$ obeys the following equation of motion:

$$u_{\tilde{k}}'' + \frac{v'}{v}u_{\tilde{k}}' + \left(\mu - \frac{a''}{a} - \frac{a'}{a}\frac{v'}{v} \right)u_{\tilde{k}} = 0, \tag{2}$$

where a denotes the scale factor and the prime denotes differentiation with respect to conformal time η. Our $u_{\tilde{k}}$ is equal to $a^2\phi_{\tilde{k}}$ from Ref. [5], while $\tilde{k}^i = a\rho^i e^{-\beta\rho^2/2}$ where ρ^i is the Fourier transform of physical coordinates x^i and

$$\mu(\eta,\rho) \equiv \frac{a^2\rho^2 +}{\left(1-\beta\rho^2\right)^2}, \quad v(\eta,\rho) \equiv \frac{e^{3/2\beta\rho^2}}{\left(1-\beta\rho^2\right)}. \tag{3}^3$$

Still not clear?

How can Greene share his work with those of us who are seriously challenged by theoretical physics and its mathematical base? In this endeavor, he might have Johannes Kepler to thank. Kepler's first book, *The Cosmographic Mystery*, asserts that the five planets known at the time fit together as the set of the five Platonic solids. I am not interested in the boring fact that Kepler was mistaken; I am interested in the brilliant way he illustrated

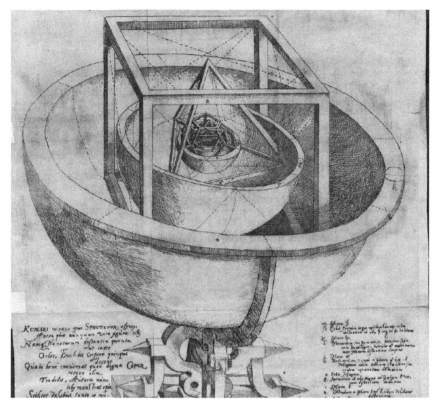

FIGURE 5.1 Kepler's five planets composed of the five Platonic solids.

his mistake in figure 5.1, not only with words but also with an image that embodies the solid geometry on which it is based.

Charles Dodgson, a mathematician, in the guise of Lewis Carroll, might easily be another predecessor. In *Alice's Adventures Wonderland,* playing cards acquire a third dimension: planes become people. Because Alice becomes frustrated with injustice at the trial of the jack of hearts, she ends it. In figure 5.2, kings and queens are transformed into a cardboard blizzard.

Like Kepler and Carroll, Greene translates abstruse mathematics into plane and solid geometry; he makes the abstract concrete, the invisible visible. At the same time, by means of visual analogies embedded in his text—references to such solid objects as balloons and Swiss cheese—he makes us feel at home, at least for a time, in the bizarre world he so comfortably inhabits. That these analogies are filched from others need not give us a moment's pause. Do we judge Shakespeare from his sources or *Henry V* from *Famous Victories*?

FIGURE 5.2 Alice ends the trial of the jack of hearts.

SOURCE: Lewis Carroll, *Alice's Adventures in Wonderland* (Boston: Lothrop, 1898), 98.

There is another point to make. Greene has written three books for adult readers—*The Elegant Universe, The Fabric of the Cosmos*, and *The Hidden Reality*. But in my view they are really one book, just as the *Forsythe Saga* is really one book. Greene's body of work is a single, extended achievement, a marvel of explanation that tells us what is true about theoretical physics and what Greene would like to be true. Settled physics is, of course, not lacking in interest. We are suitably impressed by Einstein's inventiveness in deducing that the invariance of the speed of light leads to the relativity of time and space. We have little trouble admiring how the double-slit experiment confirms the wave-particle duality of light. We are gratified to learn that the galaxies are speeding away from us at warp speed. With Greene's original contribution, however, with string theory and multiverses, there is a problem of credibility: there is no experimental or observational evidence for these theories, and there may never be. We are faced with the danger of believing six impossible things before breakfast. We are in touch with the speculative sublime, an extrapolation from current and consensual theory, brilliant enough to be arresting, bizarre enough to be incredible.

Stringing Us Along?

Some physicists feel that string theory should follow Dorothy Parker's alleged advice about a novel she was reviewing: it should not be tossed lightly aside; it should be hurled with great force. Physicist Lee Smolin says that the theory "lacks essential features necessary to describe nature."[4] Physicist Peter Woit suggests that it is "not even wrong,"[5] that it contains flaws so egregious it is not even physics. Physicist Sheldon Glashow opines that it is "a new version of Medieval theology."[6] Physicists George Ellis and Joseph Silk assert that it is "a no-man's-land between mathematics, physics and philosophy that does not truly meet the requirements of any [of these disciplines]."[7] Physicist Richard Feynman remarks, "For anything that disagrees with an experiment, [string theorists] cook up an explanation."[8] The science writer Jim Holt captures the principal argument against such allegedly vapid speculation:

> For more than a generation, physicists have been chasing a will-o'-the-wisp called string theory. . . . Yet . . . not a single new testable prediction has been made, not a single theoretical puzzle has been solved. In fact, there is no theory so far—just a set of hunches and calculations suggesting that a theory might exist. And, even if it does, this theory will come in such a bewildering number of versions that it will be of no practical use: a Theory of Nothing.[9]

The observation made by physicist John Ellis rings true today, three decades later: "There is still a yawning gap between the hopes for the superstring and its achievements."[10]

Theories of Greene's multiverse also receive stiff criticism. The venerable Freeman Dyson opines:

> String cosmology is a part of theoretical physics that has become detached from experiments. String cosmologists are free to imagine universes and multiverses, guided by intuition and aesthetic judgment alone. Their creations must be logically consistent and mathematically elegant, but they are otherwise unconstrained.[11]

Another prominent physicist, Paul Steinhardt, finds the multiverse "an unmitigated disaster as a theory. . . . Literally any physically possible cosmic property will occur in this multiverse . . . so there's no way to falsify this theory, or verify it for that matter, because anything could fit."[12]

Although Steinhardt is criticizing the inflationary multiverse, his strictures apply to any theory proposing an infinite number of universes, each with different cosmic properties. Finally, there is this from "Bob" in reply to an article on string theory "Hey, if our theories aren't validated in this universe, let's invent other universes in which they will work and use this as justification to continue receiving grant money!"[13]

Greene acknowledges that string theory and multiverses may prove dead ends, mere footnotes in the history of science. While his caution is admirable, it obscures a point important in the evolution of science, the change over time in what counts as good theory and good evidence. At one time, a good theory was expected to be deterministic, that is, to predict exactly. Einstein's theory of gravity was better than its Newtonian counterpart because, among other things, it predicted a small deviation in the orbit of Mercury over time, something Newtonian theory could not do. Then came quantum mechanics. Quantum mechanics is not classical theory because it is inherently probabilistic: it can predict with accuracy only the *likelihood* of an event. In quantum mechanics, likelihood is all there is. Still, quantum mechanics and general relativity are alike in that both make accurate predictions; both submit to the principle that "the ultimate test of one theory over another is observational success."[14]

But such a confirmation was not available in the case of the Higgs boson, whose existence is the keystone of the Standard Theory of quantum mechanics. The confidence that physicists have in the Higgs's reality and in the truth of the Standard Theory its existence confirms is entirely indirect. The boson cannot be observed, only the debris its almost immediate self-destruction leaves behind. It is from this debris that the existence of the Higgs is inferred. Physicists have established by consensus a standard such indirect proofs of existence must meet. The indirect evidence for the Higgs boson meets this standard. That is all we can say.

It is possible that string theory will meet none of these standards; it is also possible that a new standard will evolve, one in which mathematical consistency with accepted theories will be sufficient for acceptance. Should such an evolution take place, the line between science and science fiction might no longer be entirely clear; it might no longer be possible to differentiate the scientific clearly from the speculative sublime. The philosopher of science Richard Dawid makes the case that we may be witnessing the emergence of a new paradigm in what counts as science:

> Without going into detail, I note that the general scientific acceptance of specific unobservable scientific objects in the early twentieth century or the

confident prediction of new particles based on the particle physics standard model may be taken as significant stages in a process that leads toward increasing trust in theoretical conceptions based on the assessment that, under some circumstances, an equally convincing theoretical structuring of the available data with substantially different prognostics is unlikely to exist. Seen in this light, arguments from string physics do not open up an entirely new way of assessing scientific theories but merely strengthen a strand of theory assessment that has always been part of scientific reasoning. In fact, contemporary string physics does not seem to mark the endpoint of this development.[15]

But given a skepticism so deep and abiding, why should anyone pay serious attention to Brian Greene? Why should anyone read his books? Why do they do so in droves? Perhaps for the same reason they read well-written science fiction like Stanislaw Lem's *Solaris* or Ursula Le Guin's *Left Hand of Darkness*. Indeed, it is proof positive of the validity of this comparison that Greene has written *Icarus at the Edge of Time*, a work of science fiction for children of all ages, a retelling of the ancient myth of Icarus, a young man who attempts to escape from imprisonment with his father, Daedalus. According to this myth, Daedalus has fashioned two sets of wings for their aerial escape, using feathers and wax. He warns Icarus not to fly too close to the sun, as the wax will melt. But because Icarus cannot resist the temptation, he falls to his death.

In Brian Greene's retelling, Icarus is a fourteen-year-old boy on a voyage to a remote planet that has sent earth a radio signal, the sign of intelligent life in a faraway galaxy. Icarus is very bright, very sophisticated technically. When he sees an unexpected black hole approaching, he wants to investigate it in his Runabout. His father warns him not to go. But Icarus disobeys, with good reason: he has studied black holes; he knows all about them. As it turns out, he does not know everything: he does not know that the gravitational field of a black hole is so strong that time slows down dramatically. When Icarus emerges from the black hole ten thousand years later, he learns that the voyage he participated in was successful. But of course all those he knew are long dead. Greene's is science fiction based on science fact: it was Einstein who predicted that time would slow down dramatically in the gravitational field of a black hole.

Greene's book is a collaboration with Chip Kidd, Alfred Knopf's famous book designer, the creator of the cover of Jurassic Park, illustrated in figure 5.3.

FIGURE 5.3 Cover of the first edition of *Jurassic Park*.

This cover illustrates Kidd's—and Greene's—persistent guiding question: What does this story *look* like? It was sheer inspiration in the case of *Icarus* that led Kidd to the Hubble telescope photographs, arresting, appropriate, and copyright free. Figure 5.4 is an example. A selection of these represent Icarus's voyage through the cosmos. They are modified only to this extent: as the story progresses toward its climax, a pitch-black circle—a black hole—at the center of the double-fold page becomes larger and larger. As the story proceeds to its denouement, the circle diminishes in size; finally, it disappears.

In Greene's popular science books aimed at adults his subject is the same "elegant universe," a cosmos governed by a theory of everything—string

FIGURE 5.4 Galaxy NGC 1672.

SOURCE: Hubble Space Telescope website, April 3, 2007, http://www.spacetelescope.org/news/ heic0706.

theory. String theory accounts for the origin of matter at the ultra-microscopic scale, the origin and subsequent development of the universe and, possibly, a host of universes, the multiverse. Underlying all of Greene's books is an argument for the rightness of string theory, an argument based on past successes:

> Centuries of scientific investigations have shown that mathematics provides a powerful and incisive language for analyzing the universe. Indeed, the history of modern science is replete with examples in which the math made predictions that seemed counter to both intuition and experience (that the universe contains black holes, that the universe has anti-matter, that distant particles can be entangled, and so on) but which experiments and observations were ultimately able to confirm.[16]

Greene explains the history of the universe and the underlying theory that governs it by "stripping away the mathematical details in favor of metaphors, analogies, stories, and illustrations."[17] Throughout, he transforms the mathematical theories of his physics—only understood by other physicists—into words and images comprehensible to any interested reader. If, convinced by these, you accept the rightness of string theory, the resulting cosmic vision is not only as sublime and mysterious as Genesis but "thoroughly unlike what anyone ever expected."[18] Greene's is a speculative sublime, based upon formulating a set of plausible assumptions, then working out the consequences by a combination of physical theory and mathematics—no matter how apparently implausible the endpoint. It is a theory as all-encompassing as those of Feynman or Weinberg. It may even be true.

The Strings That Constitute the Universe

"Theory of everything" could mean a set of equations intended to explain "absolutely everything, from the big bang to day dreams."[19] In Greene's more restricted sense, it means instead "a theory that can explain the properties of the fundamental particles and of the forces by which they interact and influence one another."[20] Greene argues that string theory—a theory whose origins can be traced to the late 1960s—will one day satisfy that definition. In fact, he has bet his career on it. The story begins in 1984 when Michael Green and John Schwarz sparked a revolution that, according to Greene, has the potential to one day eradicate "the mathematical antagonism between general relativity and quantum mechanics, clearing a path that seemed destined to reach the unified theory."[21] Thanks to their creative thinking, a relatively obscure theory beset by mathematical problems underwent a quantum leap into prominence. Yet from the neutral technical language of the text, only theoretical physicists would have guessed that a revolution was in the offing. Here is Green and Schwarz's abstract:

> Supersymmetric ten-dimensional Yang-Mills theory coupled to $N = 1$, $D = 10$ supergravity has gauge and gravitational anomalies that can be partially cancelled by the addition of suitable local interactions. The remaining pieces of all the anomalies cancel if the gauge group is SO (32) or E8 × E8. These cancellations are automatically incorporated in the type I superstring theory based on SO (32). A superstring theory for E8 × E8 has not yet been constructed.[22]

Their article addresses and resolves the "gauge and gravitational anomalies" that had previously marred "supersymmetric ten-dimensional Yang-Mills theory coupled to $N = 1$, $D = 10$ supergravity," a low-level approximation of string theory. Green and Schwarz then eliminate mathematical anomalies in string theory itself. By doing so, they establish it as a plausible candidate for a theory of everything. Their article persuades "thousands of physicists worldwide to drop their research in progress and chase Einstein's long-sought dream of a unified theory." Greene, then a promising doctoral student at Oxford University, joins the Black Friday rush.[23]

No one has done more to make "string theory" a household name than Brian Greene, a man who pops up not only on PBS but also on *The Colbert Report.* He has become a celebrity by converting its complex mathematics and opaque jargon into a compelling story general readers can follow, experiencing thereby the echo of abstruse theory. His is not an inspirational story in the traditional sense, in which human beings are the focus. Such stories do exist in the popular science literature. Inventor John Harrison stars in Dava Sobel's *Longitude.* Naturalist Joseph Banks, astronomers William and Caroline Hershel, and chemist Humphry Davy are the central characters in Richard Holmes's *The Age of Wonder.* James Watson, Francis Crick, Rosalind Franklin, Maurice Wilkins, and Linus Pauling complete the cast in James Watson's memoir *The Double Helix.* In Greene's books, however, it is concepts, not people, who are central. In Greene's books, the history of scientific concepts is embedded in the history of science.

In the first part of Greene's master narrative, the principal problem is broached: Einstein's failure to turn the two known forces at the time—gravity and electromagnetism—into a unified theory that would resolve the paradoxes of quantum mechanics. Subsequently, it became clear to the physics community at large that this unification would require fusing Einstein's theory of general relativity, which governs the force of gravity in the macroscopic world, with quantum mechanics, which governs the three major forces at work in the microscopic world—the strong force, weak force, and electromagnetism. These two theories, however, seemed irreconcilable. In Greene's evocative prose, "The gently curving geometrical form of space emerging from general relativity is at loggerheads with the frantic, roiling microscopic behavior of the universe implied by quantum mechanics."[24]

In the second part of Greene's master narrative, string theory is born. After a dozen years of obscurity, it prospers. Progress is made in reconciling relativity and quantum mechanics when physicists view fundamental particles as massless, incredibly small vibrating strings "about a million billion times smaller than even the minute realms probed by the world's most powerful accelerators."[25] The particle associated with gravity, the graviton, is a string that vibrates one way; the particle associated with electromagnetism, the photon, is a string that vibrates another way. For string theory to work, six dimensions are required beyond the three of everyday experience. These are in the form of miniscule circular loops beyond the detection of current technology.

In the third part of Greene's master narrative a significant problem emerges. Physicists have developed five distinct and incompatible versions of string theory. Later, Ed Witten—often referred to as a modern-day Einstein—weaves these into a single framework, "M-theory." Then another problem arises. Physicists realize the number of shapes for the extra dimensions of space string theory required was on the order of 10^{500}! This problem was solved by the proposal of a "multiverse": a cosmos of 10^{500} universes to accommodate the theoretical plethora. To give us a basic understanding of the scale of 10^{500} universes, Greene essays an analogy that evokes Kant's mathematical sublime: "No familiar example of anything—not the number of cells in your body (10^{13}); not the number of seconds since the big bang (10^{18}); not the number of photons in the observable part of the universe (10^{88})—comes even remotely close to the number of universes we're contemplating."[26] 10^{500} impossible things.

Whatever the mathematical complexities of string theory, the basic concept behind it comes down to a simple analogy—it is as if the fundamental particles of nature vibrate like the strings of a harp or a violin. Greene's first figure in his first book—my figure 5.5—visualizes this analogy with a

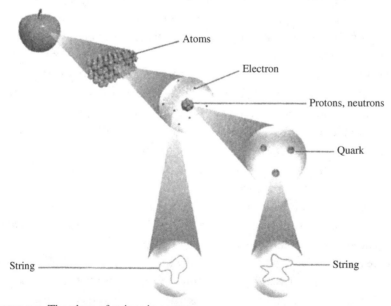

FIGURE 5.5 The place of strings in matter.

SOURCE: Brian Greene, *The Elegant Universe: Superstrings, Hidden Dimensions, and the Quest for the Ultimate Theory* (New York: W. W. Norton. 2003), 14.

sequence of linked pictures, beginning with an apple. The apple is composed of atoms, which in turn are composed of electrons, neutrons, and protons; the neutrons and protons are composed of three quarks each, and the quarks and electrons are composed of strings of different shapes. The replacement of one comparison, subatomic particles are like dimensionless points, with another, the smallest particles in nature are like one-dimensional strings or loops, is a game changer in theoretical particle physics.

The analogy is just an analogy: it helps us understand; it does not constitute understanding. Real strings have a mechanism that sets off the vibration; string theory strings do not. Real strings are observable; string theory strings are not. Real strings have mass; string theory strings do not. So what is such stuff as strings made of? Steven Weinberg has described strings as "tiny one-dimensional rips in the smooth fabric of space."[27] To Greene, they add a "new microscopic layer of a vibrating loop to the previously known progression from atoms, through protons, neutrons, electrons, and quarks."[28]

What does string theory require? First, that "empty space" is not empty: it is a scene of furious activity. That this is the case, that there are quantum field "jitters," was established as early as 1948: Dutch physicist Hendrik Casimir detected these fluctuations. He placed two metal plates in an otherwise empty region and found something extraordinary: "The quantum field jitters between the plates became a bit weaker than those outside the plates, and this imbalance *drives the plates toward each other*."[29] This is not gravitational attraction, which is far too weak to account for this effect, an effect solely a function of so-called empty space.

This is activity *within* space. Even space itself, however, is subject to quantum jitters. In figure 5.6, the uniform grid in the background represents space. As we examine this grid more closely with an imaginary magnifying glass, the smoothness eventually begins to ripple until we reach the ultra-microscopic scale, where "space becomes a seething, boiling cauldron of frenzied fluctuations." Moving from the bottom to the top of this visualization, we see the smooth space of general relativity erupt into the contorted and uncertain space of quantum mechanics.

String theory also requires more dimensions than the three of the everyday world. Is it possible to imagine dimensions we cannot experience? In 1884, Edward Abbott published *Flatland*, an intellectual journey into lives of two-dimensional creatures who live on a plane. Square, one of its inhabitants, visits Lineland, whose one-dimensional king indignantly refuses to believe that a second dimension exists. To demonstrate that it does, Square "disappears" by moving out of the king's line. The king is not enlightened but indignant (see figure 5.7).

FIGURE 5.6 The fabric of space visualized: from the massless smoothness of general relativity to the wildly fluctuating space of quantum mechanics.

FIGURE 5.7 Square vanishes from Lineland by moving out of the King's line.

Exasperated by the king's incomprehension, Square explodes in fury:

Besotted Being! You think yourself the perfection of existence, while you are in reality the most imperfect and imbecile. You profess to see, whereas you see nothing but a Point! You plume yourself on inferring the existence of a Straight Line; but I *can see* Straight Lines, and infer the existence of Angles, Triangles, Squares, Hexagons, and even Circles. Why waste more words? Suffice it that I am the completion of your incomplete self. You are a Line, but I am a Line of Lines, called in my Country a Square.[30]

Later in the book, another journey takes place: Sphere takes Square in hand and introduces him to three dimensions. If there are three dimensions, Square conjectures, why not four? Why not eight? Sphere is incredulous at Square's stubborn advocacy: "In vain did Sphere, in his voice of thunder, reiterate his commands of silence and threaten me with the direst penalties if I persisted. Nothing could stem the flood of my ecstatic aspirations. . . . I was intoxicated with the recent draughts of Truth to which he himself had introduced me."[31]

Central to string theory are its nine dimensions: the three we know firsthand plus six curled up at an ultra-microscopic scale undetectable by means of current technology. Although nine spatial dimensions seem unimaginable, Greene manages to visualize them, relying on an analogy with a garden hose. If we look at a hose from a great distance, it seems like be one-dimensional line; up close, we see its three dimensions clearly. So too the extra dimensions posited by string theory. While from our perspective the universe has only three dimensions, at the string level it has many more curled up into an origami shape. First Greene displays a single six-dimensional shape on the two-dimensional page; then in figure 5.8 he shows us a swarm of these integrated on the grid representing the usual three dimensions.

FIGURE 5.8 Six-dimensional Calabi-Yao shapes placed on grid of familiar three dimensions.

SOURCE: Brian Greene, *The Elegant Universe: Superstrings, Hidden Dimensions, and the Quest for the Ultimate Theory* (New York: W. W. Norton. 2003), 208.

Within the narrative framework of history of science, Greene employs analogies, metaphors, and images to make sense of the characters in his cosmic drama, the concepts of string theory. As a consequence, we can join Greene in experiencing a phenomenon Edmund Burke first pointed out, the dynamic and mathematical sublime of the very small, so very much smaller than Burke could possibly have imagined.

The Expanding Universe

The universe that Jack built—a universe that consists of apples that consist of molecules that consist of atoms that consist of subatomic particles that consist of quarks that consist of strings—has a long history. A deeper understanding of that history brings us to the source of Greene's deepest beliefs: "Over the last few millennia, religious and philosophical traditions worldwide have weighed in with a wealth of versions of how everything— the universe—got started. Science, too, over its long history, has tried its hand at cosmology."[32] While religion has tried, science has succeeded, insofar as success is possible:

> What I really mean to say is not that the idea [of God] is wrong, but as a scientist, I find it profoundly uninteresting, because it gives me no new insight into any of the deep questions that we've been talking about here. Doesn't help me calculate anything. Doesn't help me gain some insight into these big mysteries. It simply takes one mystery and uses another three-letter-word to re-label that mystery. And that is why I don't find it interesting. Not that it's wrong, I don't find it interesting.[33]

What *does* Greene find interesting? The tremendous progress cosmology has made and continues to make. A mere century ago, scientists believed the universe was infinite spatially, eternal, and static—neither expanding nor contracting. They believed that it consisted of the planets and stars within a Milky Way surrounded by empty space.[34] By the 1970s, the Big Bang theory had changed that picture radically. Abetted by theory, astronomical observation suggested that the universe began 10 to 20 billion years ago as an unimaginably small, dense, and hot soup; it consisted of the three building blocks of atoms—protons, neutrons, and electrons—plus a mélange of other subatomic particles: photons, neutrinos, and the antiparticles of the electron and neutrino. After an initial explosion at $t = 0$, the soup expanded and cooled, and its composition changed. At 1/100 second, the temperature was 10^{11} kelvins (that is, one followed by eleven zeros). This is unimaginably hot: the cores of the hottest stars have a temperature of 40 million kelvins—just seven zeros.

At 1/100 second after the Big Bang, for every proton there was a neutron. At one second, the temperature dropped an order of magnitude, and the balance of protons to neutrons shifted: three protons to every neutron. At three minutes, the temperature declined by another order of magnitude, and the percent of protons climbed to eighty-six. Most important for the future of the universe, atomic nuclei formed by binding some of the protons to the neutrons through nuclear fusion. Several hundred thousand years later, with continued expansion and cooling to the temperature of the sun's surface, electrons and nuclei joined to form stable atoms, primarily hydrogen and helium. About 500 million years later, as gravity caused matter to attract and gather into clumps, stars and galaxies formed from the hydrogen and helium and generated the much heavier elements that now populate the periodic table. Without this particular combination of particles and temperatures, there would be no atoms, no stars or galaxies—and no us. The latest astronomical calculations tell us that our universe is 14 billion years old, the receptacle for 100–200 billion galaxies. There may be as many as 8.8 billion habitable planets in the Milky Way alone. This is a scenario more sublime perhaps than the Alps at sunrise and as sublime perhaps as the first chapter of Genesis.

While the physics behind the origin of the universe remains mysterious, Steven Weinberg knows that

> in the beginning there was an explosion. Not an explosion like those familiar on earth . . . but an explosion which occurred simultaneously everywhere. . . . "All space" in this context may mean either all of an infinite universe, or all of a finite universe which curves back on itself like the surface of a sphere. Neither possibility is easy to comprehend, but this will not get in our way; it matters hardly at all in the early universe whether space is finite or infinite.[35]

It is Brian Greene who explains what it means for the universe to be finite or infinite, and what it means for an infinite universe to expand. To explain the finite universe "which curves back on itself like the surface of a sphere," he draws upon a balloon analogy:

> This analogy likens our three-dimensional space to the easier-to-visualize two-dimensional surface of a spherical balloon . . . that is being blown up to larger and larger size. The galaxies are represented by numerous evenly spaced pennies glued to the balloon's surface. Notice that as the balloon

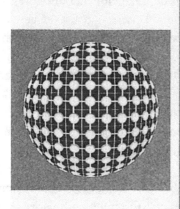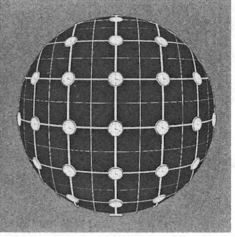

FIGURE 5.9 Analogy of expanding finite universe to expanding sphere.

SOURCE: Brian Greene, *The Fabric of the Cosmos: Space, Time, and the Texture of Reality* (New York: Vintage, 2005), 234.

expands, the pennies all move away from one another, providing a simple analogy for how expanding space drives all galaxies to separate.[36]

This analogy comes with a picture of a sphere symmetrically festooned with pennies. In figure 5.9, each image of Lincoln is facing up; each represents an observer at different locations. Greene notes that the view for any one president is the same as that for any other. Then he translates his analogy back to the actual expanding universe:

> [An] observer in any one of the universe's more than 100 billion galaxies, gazing across his or her night sky with a powerful telescope, would, on average, see an image similar to the one we see: surrounding galaxies rushing away in all directions. . . . Every point—every penny, every galaxy—is completely on a par with every other.[37]

Although the analogy works fairly well, Greene acknowledges that it is not without problems. First, while the balloon is filled with air or helium, the universe has no equivalent space. Moreover, if you look at the night sky, you will not see the equivalent of the balloon with pennies symmetrically located; you will see something more akin to Van Gogh's *Starry Night*. Greene employs another analogy to resolve this difficulty:

> Think of a glass of water. On the scale of molecules, the water is extremely heterogeneous: there is an H_2O molecule over here, and an expanse of empty

space, another H$_2$O molecule over there, and so on. But if we . . . examine the water on the "large" everyday scales we can see with the naked eye, the water in the glass looks perfectly uniform. The nonuniformity we see when gazing skyward is like the microscopic view from a single H$_2$O molecule.[38]

There is another problem: the universe might not be curved like a sphere but flat. To visualize the expansion of this sort of universe, Greene superimposes a rectangular symmetrical grid in which our pennies are positioned at the nexuses. As the universe expands and the pennies move apart, the whole remains symmetrical, just as in the balloon analogy. But there is still a problem, because a rectangle has borders. Greene's solution is to modify the analogy: a rectangle with borders like a tabletop is transformed into a flat video game screen in which animations move off an edge on one side only to immediately reappear on the other.

Figure 5.10a represents this finite universe. The galaxy at the top left is leaving the rectangle and reappearing at the upper right; at a later time, the large galaxy at the lower left will leave the screen and reappear on the right. In figure 5.10b, time advances from left back to right front. In it, we see time-space slices of this finite universe starting from the time represented in figure 5.10a (right front) and traveling backwards in time toward the Big Bang. We can easily imagine the time-spaces slices contracting as any celestial objects within come closer and closer together until they converge into an infinitesimal point just prior to the Big Bang.

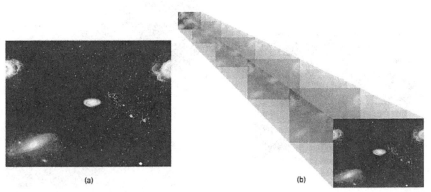

(a)   (b)

FIGURE 5.10 Analogy of flat and finite expanding universe to a rectangular video game screen that expands over time.

SOURCE: Brian Greene, *The Fabric of the Cosmos: Space, Time, and the Texture of Reality* (New York: Vintage, 2005), 244.

Let us suppose the universe is not finite but infinite. What can it mean for an infinite universe to expand? At first glance, the question is nonsense. But Greene begs to differ:

> If space is infinite and you shrink all distances by a factor of two, the size of space becomes half of infinity, and that is still infinite. So although everything gets closer and the densities get even higher as you head further back in time, the overall size of the universe stays infinite.[39]

In figure 5.11a, we imagine the universe as an infinite plane populated by galaxies; in figure 5.11b, we go back in time. Although the planes remain infinite, the galaxies become progressively denser. Finally, at t = 0, they reach infinite density. Figure 5.11 differs from figure 5.10 in that the size of the rectangle does not change as we retreat in time—half of infinity is still infinity. In this scenario, "the big bang eruption took place everywhere on the infinite expanse" rather than at an infinitely dense point. However far-fetched, this is nevertheless "the front-running contender for the large-scale structure of spacetime."[40]

When did this expansion start? In one string theory scenario—a scenario that staggers the imagination—13.82 billion years ago the universe was compressed into a single string. At the time of the Big Bang, all its nine dimensions were curled up into that tiny space. At the "Planck time" of 10^{-43} seconds, three of these expanded by "a factor larger than a million

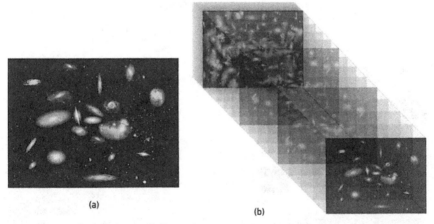

(a)

(b)

FIGURE 5.11 An infinite expanding universe remains the same size but becomes less cluttered over time.

SOURCE: Brian Greene, *The Fabric of the Cosmos: Space, Time, and the Texture of Reality* (New York: Vintage, 2005), 249.

trillion trillion in less than a millionth of a trillionth or a trillionth of a second."[41] The other dimensions remained string-sized. Also at the Planck time, gravity parted ways with the other three forces: the strong force, the weak force, and the electromagnetic force. The "grand unification" of the three non-gravitational forces lasted until 10^{-35} seconds, at which point the strong force separated from its weak and electromagnetic counterparts. In the period of "electroweak unification," which lasted until 10^{-12} seconds had passed, the weak and electromagnetic forces separated.

To better illustrate the physics of these phase transitions, Greene employs another aquatic analogy. He asks of us to think of a glass of water:

> The molecules of H_2O are uniformly spread throughout the container and regardless of the angle form which you view it, the water looks the same. Now watch the container as you lower the temperature. At first not much happens. On microscopic scales, the average speed of the water molecules decreases, but that's about all. When you decrease the temperature to 0 degrees Celsius, however, you suddenly see that something drastic occurs. The liquid water begins to freeze and turn into solid ice.[42]

During the electroweak unification period, the theory also predicts that the Higgs field forms and fills all space, giving mass to elementary particles by slowing their motion "much as a vat of molasses resists the motion of a Ping-Pong ball that's been submerged."[43] In another analogy, Greene compares particles to well-known personalities:

> Photons zip through the Higgs ocean [at the speed of light] as easily as B-movie has-beens slip through the paparazzi, and therefore remain massless. W and Z particles though, like Bill Clinton and Madonna, have to slog their way through acquiring masses that are 86 and 97 times that of a proton, respectively.[44]

In figure 5.12, Greene visualizes this version of the Big Bang. At a glance, we see these key transitions as the early universe expands and its temperature plummets by sixteen orders of magnitude—from 10^{32} to 10^{16} kelvins in 10^{-12} seconds. We also see over time the progressive formation of atoms, galaxies and, finally, our solar system. While these are formed, the temperature of the universe drops over ten billion years by the same order of magnitude as it did over the first 10^{-12} seconds.

In figure 5.13 Greene depicts the expanding universe from t = 0 to the present. The point at the far left of represents the Big Bang; the even sides,

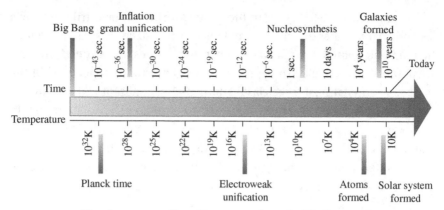

FIGURE 5.12 Time/temperature line of key events from the Big Bang to the present.

SOURCE: Brian Greene, *The Elegant Universe: Superstrings, Hidden Dimensions, and the Quest for the Ultimate Theory* (New York: W. W. Norton. 2003), 356.

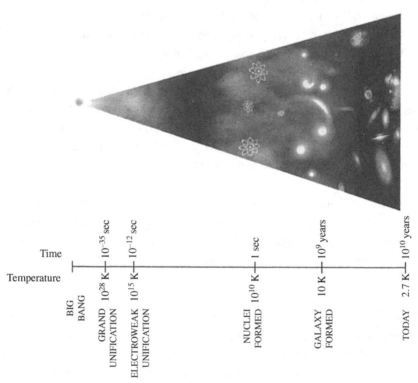

FIGURE 5.13 Analogy of finite expanding universe to isosceles triangle with timeline below.

SOURCE: Brian Greene, *The Fabric of the Cosmos: Space, Time, and the Texture of Reality* (New York: Vintage, 2005), 270.

the uniform expansion to the present. Within the triangle are representations of the formation of atomic nuclei one second after the Big Bang and of the first galaxies at 10^9 years. Greene can discuss with "some confidence" only the unification at 10^{-12} seconds of electromagnetism and the weak force: only it has been confirmed experimentally.

The triangle in figure 5.13 reflects expansion of the universe at a uniform rate. However, it now seems clear that the rate of expansion was not uniform. The mechanism for this inflation was gravity acting as a repulsive rather than attractive force. The rate of expansion also underwent a prolonged burst after the first seven billion years because the effects of a form of energy, dark energy, surpassed those of gravity. Greene represents this variable expansion in a modified version of figure 5.12 shaped like a bell.[45]

Greene's origin story is certainly evocative of wonder, but also of wondering: Can this fantastic story possibly be true? But it is our next topic, the multiverse, that stretches our imaginative resources almost past the breaking point.

The Multiverse

Is our universe the only one? Why couldn't there be a numberless number of universes within the infinite universe depicted in figure 5.11b? Figure 5.14 shows us that each universe would have its own cosmic horizon beyond which no one within could ever observe. And in one of those infinite universes, Greene imagines that

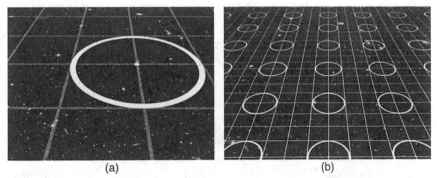

(a)   (b)

FIGURE 5.14 *Left*, single universe like ours within infinite space. *Right*, infinite multiverse composed of a numberless number of individual universes.

SOURCE: Brian Greene, *The Hidden Reality: Parallel Universes and the Deep Laws of the Cosmos* (New York: Vintage, 2011), 33.

there's a galaxy that looks like the Milky Way, with a solar system that's the spitting image of ours, with a planet that's a dead ringer for earth, with a house that's indistinguishable from yours, inhabited by someone who looks like you, who is right now reading this very book and imagining you, in a distant galaxy, just reaching the end of this sentence.[46]

Suppose the inflationary period did occur in our universe. Why would such an event happen only once? Isn't it possible that multiple universes have been born with a bang like the Big Bang? Isn't it possible that these universes underwent a similar inflationary period sparked by changes within what Greene terms the "inflation field" that permeates the multiverse? Here is his way of visualizing this phenomenon:

> Think of the universe as a gigantic block of Swiss cheese, with the cheesy parts being regions where the inflation's field value is high and the holes being regions where it's low. That is, the holes are regions, like ours, that have transitioned out of the superfast expansion and, in the process, converted the inflation field's energy into a bath of particles, which over time may coalesce into galaxies, stars, and planets. In this language, we've found that the cosmic cheese acquires more and more holes because quantum processes knock the inflation's value downward at a random assortment of locations. At the same time, the cheesy parts stretch ever larger because they're subject to inflationary expansion driven by the high inflation field value they harbor. Taken together, the two processes yield an ever-expanding block of cosmic cheese riddled with an ever-growing number of holes.[47]

Another candidate for the multiverse emerges from string theory, which allows for shapes other than one-dimensional strings: membranes that can have up to nine dimensions. A point is a zero-dimensional membrane, while a string is its one-dimensional counterpart. In string theory, our universe is a three-dimensional brane floating within a vast expanse of extra dimensions we cannot observe because the light we see is confined to our three-brane universe. An analogy helps: the universe is "an extraordinarily thin slice of bread" interspersed with "the Orion, Horsehead, and Crab nebulae; the entire Milky Way; the Andromeda, Sombrero, and Whirlpool Galaxies; and so on—everything within our three dimensional spatial expanse, however distant."[48] In figure 5.15, Greene sees parallel universes as extraordinarily thin slices of bread, some like ours; some, perhaps, very different. These could be a very short distance away in another dimension from which we are barred, just as the king of Lineland was barred from Flatland.

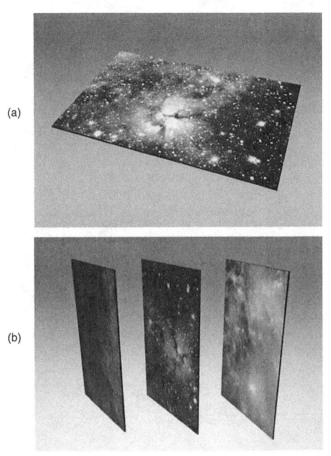

(a)

(b)

FIGURE 5.15 *Top*, two-dimensional representation of three-brane like our universe. *Bottom*, three parallel three-branes.

In an early version of the Big Bang, the universe expanded until the force of gravity slowed expansion to a standstill. Then, as gravity continued to act, over time the universe contracted to an infinitely dense point, a process that occurred again and again. Greene's theory offers an alternative to this picture, illustrated in figure 5.16. This other cosmic history begins with the collision of two three-branes separated by a short interval: "The universe of which we are directly aware corresponds to one of these three-branes; if you like, you can think of the second three-brane as another universe. . . . . It's hovering no more than a fraction of a millimeter away (the separation being in the fourth spatial dimension . . . , but because our three-brane is so sticky and the gravity we experience so weak, we have no direct evidence of its existence."[49]

At stage 1, the two three-branes rebound from the cosmic collision illustrated at the top of the figure. At stage 2, the branes undergo

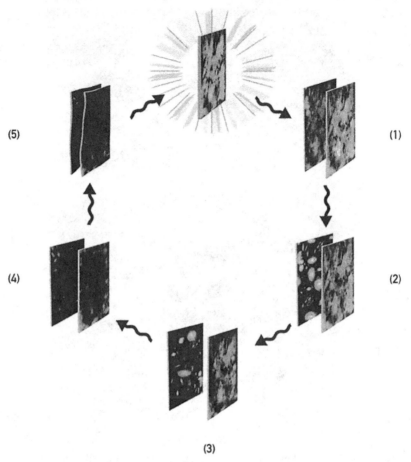

FIGURE 5.16 Stages in few-trillion-year cycle involving collision, rebound, attraction, and another collision with two parallel three-branes.

SOURCE: Brian Greene, *The Fabric of the Cosmos: Space, Time, and the Texture of Reality* (New York: Vintage, 2005), 408.

hyperexpansion and cooling, followed by a slowing of the expansion due to gravity. At this point, stars and galaxies form. At stage 3 the galaxies move farther and farther apart and the effects of gravity diminish to the point where the mysterious force, "dark energy," rules. Stage 4 kicks in some trillion years later, when the branes have expanded to a point where "matter and radiation are diluted almost completely away." At stage 5, as the two branes approach each other, "quantum jitters" cause ripples in space. At this point, the branes collide and the cycle renews. This vision reveals to us Kant's mathematical and dynamic sublime rolled into one: we experience at second hand the vastness and power of an imagined world.

An imagined world that, Greene informs us, cannot be the way things are. Entropy, the tendency of any closed process eventually to run down, bars eternal cycling. So does quantum mechanics: "Quantum mechanics ensures that there is always a nonzero probability that a chance fluctuation will disrupt the cyclic process (e. g. one brane twists relative to the other), causing the model to grind to a halt. Even if the probability is minuscule, sooner or later it will surely come to pass."[50]

Conclusion

Greene cautions that "the braneworld scenario and cyclic cosmological model it spawned are highly speculative. I have discussed them here not so much because I feel certain that they are correct, as because I want to illustrate the striking new ways of thinking about the space we inhabit and the evolution it has experienced that have been inspired by string/M-theory."[51] Such caution is not isolated. Greene frequently hedges in his explanations of string theory and other multiverses; indeed, throughout his books, he assumes his readers are highly skeptical about his Alice-in-Wonderland claims of nine spatial dimensions, the braneworld, and multiple universes. Greene is also conscientious in differentiating between what is known with great assurance and what is primarily speculative. For careful readers attuned to this distinction, the experience of the scientific sublime arises from wonder at the revelation not of how the universe *works*, but of how it *may possibly work*, a vision that evolves vividly before these readers, a product of Greene's literary skill.

CHAPTER 6 | Stephen Hawking

The Scientific Sublime Embodied

Life seemed to him a gift; the statement "I am alive" seemed to him
to contain a satisfactory certainty and many other things, held up as
indubitable, seemed to him uncertain.

—JAMES JOYCE, *Stephen Hero*

LUCY HAWKING HAS had the good fortune of being the daughter of the
most famous living physicist; she has had the better fortune of having
been a teenager before Stephen Hawking became famous, a time when he
was known and respected only by other theoretical physicists. In this less
hectic time, he was just a father, a man with a disability, to be sure, but
not a disabled man, a sufferer from Lou Gehrig's disease who defied the
odds. Who could view as disabled a man who zipped through the streets
of Cambridge in a Formula 1 electric wheelchair driven at reckless speeds
and, on one occasion at least, almost disastrous consequences? Hawking
is now, perhaps, the most famous physicist since Einstein. While his work
significantly expands the territory of the scientific sublime, his life embodies that sublime. This is not the ethical sublime that Rachel Carson, the
subject of the next chapter, embodies; it is not a code of conduct. Rather,
it is our firm sense that we are dealing with an extraordinary human being
who has overcome daunting challenges to become an impressive virtual
presence, a man who, alone among contemporary scientists, is a star, nay,
a superstar. Confined to a wheelchair, he towers above us, an exemplar,
a demonstration of just how deep a deep-seated commitment to science
can be.

But is he any good at physics? Is it all hype? His heroes—Galileo, Newton, and Einstein—are models he cannot hope to emulate. Those on whom he consistently relies—Werner Heisenberg, Paul Dirac, and Richard Feynman—are clearly his superiors. True, he is an elite physicist honored by his peers, but he is more a Dom than a Joe DiMaggio, excellent, though not the very best. As he says himself, "To my colleagues am just another physicist." But his professional reputation hardly matters, because, as he asserts with characteristic good humor:

> To the wider public I became possibly the best-known scientist in the world. This is partly because scientists, apart from Einstein, are not widely known rock stars, and partly because I fit the stereotype of a disabled genius. I can't disguise myself with a wig and dark glasses—the wheelchair gives me away.[1]

While Richard Feynman's life in physics and his life in life are equally exercises in self-fashioning, differentiating the one from the other is possible; this is the work of Feynman's talented biographer, James Gleick. Feynman knows the difference as well and is scrupulous in honoring it: his popular works of science infuse textbook accounts with life, but not with his life. Stephen Hawking is of another mind altogether, a physicist without borders, depicted and yet not quite depicted in his autobiography, in biographies, in memoirs by his ex-wife, and on film, many films, a myth of genius and indomitable will, a man who has handicapped his handicap.

How can we make sense of a scientist who has made the universe his universe, fusing literary imagination and mathematical physics into a single, powerful public message? Because Hawking's phenomenal success as a writer of popular science must not be allowed obscure the seriousness of his educational mission, it is entirely appropriate that the exploration of his work begins with a series of educational books for children, written in collaboration with Lucy, his novelist daughter.

The Hawkings' Secret Key to the Universe

When a minor novelist teams up with famous physicist, it does the book's sales no harm. Last time I looked, Lucy's first novel, *Jaded*, was at 2,755,994 on Amazon sales listing; *George's Secret Key to the Universe* was 11,312. I do not mean to imply that this children's book does not succeed on its own merits; it does. George, our story's hero, longs for a computer his parents deny him: they abhor all modern technology. A computer is available, but only as a prize for the best speech about science, a competition among elementary schools. Ignorant of science, George has no hope of winning, no hope, that is, until he meets Annie, the girl next door, and her father, Eric, a theoretical physicist armed with a supercomputer, Cosmos, a miraculous device that turns its users into space tourists. This is just what George and Annie become, circling the universe perched on a comet, experiencing the mathematical and the dynamic sublime at first-hand.

There is another physicist in the picture, George Reeper, once Eric's friend and colleague, now his bitter enemy. Reeper steals Cosmos. Then he tricks Eric into engaging in a wild goose chase in space searching for a habitable planet. In this search, Eric falls into a black hole from which, it is thought, nothing can escape. George and Annie manage to repossess Cosmos, but George is stumped in his rescue attempt; he knows nothing about black holes, and Eric's book on the subject is beyond him. Luckily, some loose pages flutter down; on them is a simple explanation of the nature of black holes, the secret key to Eric's salvation. Eric is rescued, except that he now sports someone else's eyeglasses. Cosmos has reconstituted him from the information that escaped the black hole just as his incomprehensible book predicted:

> "Slow down, Annie, slow down," said Eric, who seemed rather dazed. "You mean I've been inside a black hole and come back again? But that's incredible! That means I've got it right—that means all the work I've done on black holes is correct. Information that goes into a black hole is *not* lost forever—I know that now![2]

Armed with the experience gained from Cosmos and a self-confidence markedly increased by rescuing both Cosmos and Eric, George goes on to win the speech contest and claim the prize.

Some may object to this mixture of science and fantasy: computers don't open portals into space, boys and girls don't ride on comets to tour space, and computers can't reconstitute Hawking radiation into the configuration of matter that generated it. But that is not the point. Fantasy is vehicle: it leads to the real science of the cosmos, the authentic experience of the mathematical and dynamic sublime. Physics, it turns out, is not some dry-as-dust subject. As George puts it in his winning speech, "Some people think science is boring, some people think it's dangerous—and if we don't get interested in science and learn about it and use it properly, then maybe it *is* those things. But if you try to understand it, it's fascinating, and it matters to us and to the future of our planet."[3]

George has experienced this fascination, this sense of wonder, during his earlier trip through the universe:

> Behind the comet was a tail of ice and dust, which was getting steadily longer. As it grew, it caught the light from the faraway Sun and glistened in the wake of the comet, making it look as though thousands of diamonds were shining in outer space.
> "That's beautiful," whispered George.[4]

To be sure, George's vocabulary is not yet adequate to his experience; still, his whisper indicates his awe at what no human being has ever seen close up, a depiction science and only science can reconstruct for his delectation and our own. Another, deeper sense of wonder comes from George's rescue of Eric, a maneuver that depends on his understanding of black holes. While the quantum mechanics behind this explanation is beyond George, some form of explanation is not: "Tiny fluctuations in space and time," he reads, "meant that black holes couldn't be the perfect traps they were once thought; instead they would slowly leak particles in the form of Hawking Radiation."[5]

As George grows older, he may come to understand black holes better; if he has a talent for physics, he may contribute to their further understanding. But if his knowledge of science remains rudimentary, it may be the case not that George has failed but that his teachers have failed to make science interesting enough to capture and keep his attention as a constant source of wonder:

> In schools science is often presented in a dry and uninteresting manner. Children learn it by rote to pass exams, and they don't see its relevance to the world around them. Moreover, science is often taught in terms of equations. Although equations are a precise and accurate way of describing mathematical ideas, they frighten most people. When I wrote a popular book recently, I was advised that each equation I included would halve the sales. I included one equation, Einstein's famous equation, $E=mc^2$. Maybe I would have sold twice as many copies without it.[6]

The man who collaborated on a series of books about a young boy experiencing science first-hand was no literary novice. He had recently written an adult version of the George books, a blockbuster called *A Brief*

History of Time. Readers enter this book to begin a journey into the cosmos accompanied by a knowing guide with a sense of humor. At the same time, they find themselves traveling into the mind of a world-famous physicist, a grand theorist of black holes. To read A *Brief History of Time* is not merely to learn about science; it is to experience it through the eyes of a talented physicist.

From Black Holes to the Big Bang

A Brief History of Time is a publishing phenomenon: at least ten million copies have been sold to date; the book has been translated into at least thirty languages. No one in history has brought complex science before such a vast reading audience. About this runaway success, Hawking says: "I have sold more books on physics than Madonna has on sex." Beyond the humor sprinkled throughout the book to better engage readers, of course, is a much more serious educational mission. In *Brief History*, without having to puzzle over any equations except $e = mc^2$, readers learn about the past of cosmology from Aristotle to Newton, from Newton to Einstein, from Einstein to Dirac and Heisenberg. As they do, as the universe becomes unimaginably large, they are aided in their exploration of the mathematical and dynamic sublime it embodies. They travel to distant galaxies, to radiation left over from the Big Bang, and to black holes, the collapsed remains of dead stars; they go back in time, visiting the beginning of the cosmos, the Big Bang itself. They move outward to a distant future.

Hawking opens *A Brief History* with an unsolved puzzle regarding our two best theories of matter, both children of the 20th century, Einstein's general relativity and quantum mechanics, a theory of the very large and a theory of the very small. Both are needed if the origin of the universe is to be explained. But prior to this explanation readers must learn what a scientific theory is:

> In order to talk about the nature of the universe and to discuss questions such as whether it has a beginning and an end, you have to be clear about what a scientific theory is. I shall take the simple-minded view that a theory is just a model of the universe, or a restricted part of it, and a set of rules that relate quantities in the model to observations that we make. It exists only in our minds and does not have any other reality (whatever that might mean). A theory is a good theory if it satisfies two requirements: It must accurately describe a large class of observations on the basis of a model that contains

only a few arbitrary elements, and it must make definite predictions about the results of future observations.[7]

Hawking informs us that both general relativity and quantum mechanics have many times over passed the confirmation test of "definite predictions about the results of future observations." Yet there is a difficulty with these two very best theories: each is incomplete. According to our current understanding, there are four forces in the universe: the strongest is the strong force, which confines quarks into protons and neutrons and binds those protons and neutrons together in atomic nuclei; then there is the weaker electromagnetic force, which holds electrons around those nuclei, and is responsible for magnetic fields and electricity and light; then there is the phenomenally weaker weak force, responsible for tiny particles called neutrinos that appear in nuclear decays. Far, far weaker than any of these forces is gravity, the attractive force that depends on an object's mass. Physicists have succeeded in describing the first three forces quantum mechanically in the Standard Model. But as yet there is no theory encompassing all four forces. Likewise, general relativity, which describes gravity so well, has no room for quantum mechanics. Both these theories will fail in places where both gravity and the three forces of the Standard Model are important: the inside of black holes and the very early moments of the universe.

The two best theories are not only incomplete; they are incompatible. Relativity is a classic theory in which it is possible to accurately measure the position and velocity of particles; quantum mechanics is a theory in which this feat is impossible. This difficulty may not seem important in cosmology, which is, after all, about the very large, the domain of general relativity. There Einstein's model performs very well; it is a good theory according to Hawking's criterion. But the very large was once very small; the universe is believed to have started as a point of infinite density called a "singularity." Yes, I know that a point is dimensionless, and that infinite density makes no sense in our world of middle-sized objects. But I am not making this up, and neither are cosmologists. Using the tools of mathematical physics, they infer this initial condition from the current state of the universe. This is the best that the best physics can do.

While many have tried, no one has managed to combine quantum mechanics and general relativity into a comprehensive theory that has reached the stage of consensual science among physicists. However, one man, Stephen Hawking, has combined them in a particular case, black holes. How is a black hole formed? When a star that is sufficiently large, a star much larger than the sun, runs out of fuel, gravity takes over and it collapses; the star becomes so small and dense that even light cannot escape. Since nothing can exceed the speed of light, nothing can escape. Although we can't see black holes—they are black, remember—we can depict them according to theory. Depicted in figure 6.1 is a black hole about to ingest anything in the vicinity of its edge, its event horizon: stars, planets, electromagnetic and gravity waves, interstellar gas and dust. The black hole is the ultimate cosmic Cuisinart. No matter what falls into a black hole, the only things that will change are the total mass, the total electric charge, and the speed at which the invisible black hole rotates.

Because black holes are invisible, theory had to come first: "Black holes are one of only a very small number of cases in the history of science in which a theory was developed in great detail as a mathematical model before there was any evidence from observations that it was correct."[8] Hawking traces the origin of that theory to 18th-century England, when a talented member of the Royal Society, John Mitchell, shared with

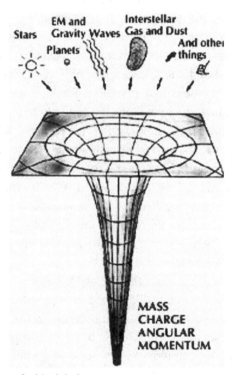

FIGURE 6.1 Diagram of a black hole.

SOURCE: Interview with John A. Wheeler, *Cosmic Search*, http://www.bigear.org/vol1no4/wheeler. htm.

his colleagues his insight that a celestial body of sufficient density would exert a gravitational force so strong that light could not escape from it. Such a body might be detected by the influence of its gravitational force on adjacent bodies:

> If there should really exist in nature any bodies whose density is not less than that of the sun, and whose diameters are more than 500 times the diameter of the sun, since their light could not arrive at us; or if there should exist any other bodies of a somewhat smaller size, which are not naturally luminous; of the existence of bodies under either of these circumstances, we could have no information from sight; yet, if any other luminous bodies should happen to revolve about them we might still perhaps from the motions of these revolving bodies infer the existence of the central ones with some degree of probability, as this might afford a clue to some of the apparent irregularities of the revolving bodies which would not be easily explicable on any other hypothesis.[9]

No one paid attention to Mitchell's speculation, perhaps because no 18th-century telescope could confirm it. It was not until 1928 that a young Indian graduate student, Subrahmanyan Chandrasekhar, a man whose Nobel Prize was many years in the future, suggested to his advisor, Arthur Eddington, that a dying star above a certain mass might eventually collapse, forming a tiny object of immense gravitational force. Eddington was discouraging; he thought the idea too unlikely to pursue. In the Soviet Union about the same time, Lev Davidovich Landau, who knew nothing of either Mitchell or Chandrasekhar, was struck by the same idea. Again, no one paid attention. In 1939, two Americans, J. Robert Oppenheimer and his student Hartland Snyder, published "On Continued Gravitational Attraction," a paper that worked out the mathematics of black hole collapse. Given the outbreak of World War II in the same year, few paid much attention. Besides, there were "no observational consequences that could be detected by the telescopes of the day."[10]

Hawking notes that the detection of black holes is a "bit like looking for a black cat in a coal cellar."[11] This particular black cat became partially visible in the 1960s with "a great increase in the number and range of astronomical observations brought about by the application of modern technology."[12] Hawking summarizes several of these. In 1963, Maarten Schmidt discovered a faint starlike object emitting an amount of energy consistent with a black hole. In 1967, a Cambridge graduate student, Jocelyn Bell-Burnell, discovered "neutron stars," whose small size and density are close to that of a star collapsing into a black hole. In the 1970s, astronomers discovered that Cygnus X-1 is a star orbiting around something no one can see, something that is likely a black hole. More recently, the Hubble space telescope captured deep within elliptical galaxy M87 a disk of gas 130 light years across. It is rotating around a black hole two thousand million times the mass of the sun.

Coincident with these astronomical observations, the theory of black holes advanced with the emergence of an international group of stellar physicists: Werner Israel, a Canadian; Roy Kerr, a New Zealander; Neil Porter and Trevor Weekes, two Irishmen; Jacob Bekenstein, an Israeli; Wolfgang Pauli, an Austrian; and Roger Penrose, an Englishman. Nevertheless, we learn from *A Brief History* that there is a central figure, Stephen Hawking. First and foremost, working with Penrose between 1965 and 1970, he made the remarkable prediction that

> there must be a singularity of infinite density and space-time curvature within a black hole. This is rather like the big bang at the beginning of time. . . . At this singularity the laws of science and our ability to predict the future would break down. However, any observer who remained outside the black hole would not be affected by the failure of predictability, because neither light nor any other signal could reach him from the singularity.[13]

Hawking's second career-making conjecture was that black holes are not as black—not as inert—as had been thought. His model predicted that black holes have a temperature; they radiate energy as particles leak out. As a consequence, they will eventually shrink in size until they vanish.

The story of the black hole tells us something about the scientific sublime that Adam Smith missed: a result can be so astonishing that the scientists themselves refuse to believe it. Hawking himself is a victim of this sensible incredulity. Even though his mathematics has shown him that black holes must radiate energy, he can't believe it. He searches in vain for a flaw in his reasoning.[14] When he presents his results at a scientific meeting, he is greeted with the same sensible incredulity on the part of the chair, John G. Taylor, who

> claimed that it was all nonsense. He even wrote a paper to that effect. However, in the end most people, including John Taylor, have come to the conclusion that black holes must radiate like hot bodies if our other ideas about general relativity and quantum mechanics are correct. Thus even though we have not yet managed to find a primordial black hole, there is

fairly general agreement that if we did, it would have to be emitting a lot of gamma rays and X rays.[15]

What went wrong in this case? Nothing. *A Brief History* tells us that resistance to new ideas is sometimes essential if science is to turn out right in the end. Opposition forces its targets to strengthen their arguments, to perfect their proofs. Hawking was right to stick to his guns; so was Taylor.

Black holes were also Hawking's key to the origin of the universe. It was Hawking who proposed early on that it began as a singularity, a dimensionless point—size zero; its subsequent expansion was the exact reverse of a black hole collapsing into a singularity. At this dimensionless initial point, no explanation for expansion could be offered because the laws of physics no longer applied. In *A Brief History*, Hawking speculates on an alternative origin story, one with no boundary in time and therefore with no need for explanatory despair. To grasp this alternative, we picture the universe as a globe like the earth. According to this analogy, "Even though the universe would have zero size at the North and South Poles, these points would not be singularities, any more than the North and South Poles are singular[ities]. The laws of science will hold at them, just as they do at the North and South Poles on earth."[16]

During the expansion of the universe, time would have three metaphorical arrows: the psychological, our perception of time passing; the thermodynamic, the movement from a state of order to disorder; and the cosmic, the rapidly expanding universe itself. But because time passing is just our perception of increased disorder, Hawking feels that the psychological arrow can be dropped from consideration. He argues that the best explanation for the two remaining arrows is his no-boundary theory. According to this, the universe has expanded from a highly ordered state into ever greater disorder, an expansion in which the cosmic and thermodynamic arrows were perfectly in synch. This is entirely a good thing. If time's two arrows did not move forward together as they do, we would not be around to observe them. This is the weak anthropic principle: only in a universe that could eventually support life would there be living beings capable of observing and reflecting upon the fine tuning that made life possible.

Physicists might beg to differ. First, they might point that the no-boundary theory is sheer speculation. Second, they might object that the weak anthropic principle is trivially true, that it says only that we are here because we are here. Third, they might think it naive to equate our perception of time with the thermodynamic arrow. The sentence "This is happening now" is always false, but this is hardly a consequence of thermodynamics. Lastly, they might object to the importance Hawking gives to the value of his discoveries and to his role as a discoverer. None of these objections prevent general readers from being swept off their feet into the speculative sublime of black holes and the origin of the cosmos; none prevent *A Brief History* from being a bestseller.

The Physics of Time Travel

The Universe in a Nutshell gives us insight into Hawking's way of doing science. Among other topics he discusses time travel—whether the laws of physics permit it. Hawking's answer is no, to the disappointment of science fiction fans everywhere. But because his arguments belong to mathematical physics, he is reduced to saying "I can show that"[17] without actually showing, and "I can also prove"[18] without actually proving. But while he fails to give us an intuitive understanding of the arguments that convinced him, he succeeds in conveying an intuitive understanding of how he transforms scattered bits of theory from wildly incompatible realms into successive probes, each of which constitutes a step forward in answer to the central question: Can the universe be so warped by the forces within it that time doubles back on itself? Can Marty McFly, accidently transported into the past, make sure his parents marry, so that he can safely go back to the future?

To follow Hawking in this quest—to join him in his own private stratosphere—is to be astonished by a remarkable display of intellectual agility. His initial attempt at a solution involves general relativity, a theory in which the masses in the universe—its stars, planets, galaxies—warp space-time, a distortion we call gravity. Only if space-time is warped in this way is a path to the past possible. But that path is blocked. Because the universe of general relativity contains only positive energy, no solution within its framework can work. To make time travel possible, negative energy is needed; quantum theory is needed. Within quantum theory, however, there is no way to explain the warping of space-time; there is no quantum theory of gravity. We need *both* general relativity and quantum theory to discover whether time can loop back on itself. But general relativity and quantum theory are incompatible.

Or are they? At this point, Hawking retrieves his earlier work on black holes. These radiate because a particle with positive energy escapes when an anti-particle of negative energy is captured by the hole. As a consequence of this radiation, black holes slowly evaporate:

> The evaporation of black holes shows that on the quantum level the energy can sometimes be negative and warp space-time in the direction that would be needed to build a time machine. Thus we might imagine some very advanced civilization could arrange things so that the energy density is sufficiently negative to form a time machine that could be used by macroscopic objects such as spaceships.[19]

At this point Hawking looks at his problem through another lens: Richard Feynman's idea that a quantum particle's path can be determined by summing up all its possible paths. In some of the paths, a particle will travel back in time:

> It seems, therefore, that quantum theory allows time travel on a microscopic scale. However, this is not much use for science fiction purposes, such as going back and killing your grandfather. The question therefore is: can the probability in the sum over histories be peaked around spacetimes with macroscopic time loops?[20]

Hawking's answer relies on an inference drawn from the Einstein universe, a discarded theoretical model in which the cosmos is finite and static. This model is useful to Hawking not because it is correct but because it is mathematically equivalent to other, far less tractable models in which time loops to the past are at least possible. Hawking pictures an Einstein universe rotating around its axis. As in a carousel, the further from the center you are, the faster you go, up to a limit: "the critical rate of rotation below which no part of the universe is rotating faster than light."[21] From the analogy with the Einstein universe, Hawking infers that

> the probability of having sufficient warping for a time machine is zero. This supports what I have called the Chronology Protection Conjecture: that the laws of physics conspire to prevent time travel by macroscopic objects.[22]

As a consequence, time loops that make time travel for us possible are, if not entirely out of the question, unlikely.

We are now at the end of our methodological journey, a voyage that corresponds to no known schematization of the scientific method. Hawking hops gingerly from theory to theory to support a conclusion for which no empirical evidence can be offered—or can ever be offered. Moreover, he is supremely indifferent to the fact that his conclusion offers no support whatever for the validity of any of the theories used to attain it. The process shows us the theorist as *bricaleur*, a tinkerer extraordinaire like Feynman, not a grand architect like Newton or Einstein.

A Grand Design Absent a Designer

When Richard Feynman said that no one understands quantum physics, he did not mean that no one could become a theoretical physicist. He meant that at every turn the physics of the micro-world violates common sense, one of whose primary dictates is that we can be certain where anything is, and if it is moving, we can calculate how fast it is going. You are in your car just past exit 45B on I-275 when you are going sixty miles an hour. How do you know? You look at the exit sign and, at the same time, you read the speedometer. The car and the freeway are made up of elementary particles such as electrons, right? So it seems to make sense that you can simultaneously measure the position and velocity of one of electrons that makes up the car you are driving. But when you try to measure the velocity and position of that electron at an instant of time, just as you did with your automobile, you encounter a principle that reigns in the micro-world—uncertainty. While in *The Grand Design* Hawking and his co-author, the Caltech physicist Leonard Mlodinow, cannot resolve this incompatibility, they can give us an intuitive understanding of a fact that defies common sense. They can show us that, in this case at least, there is a grand design.

There are laws in the micro-world. They are associated not with certainties but with probabilities, with the *likelihood* that an electron will take one path or another. These are fundamental, probabilities built right into the micro-world. They deviate markedly from our common-sense ideas of probability. Let's say you are playing darts. Based on your past performance, your companions predict that you are unlikely to hit the bullseye with your next throw. This is just an informed guess, but in principle their prediction could be made more accurate. Given the point where the dart leaves your hand and its initial speed, spin, and trajectory, given its weight and its shape, given the air resistance and the force of gravity, given the distance from the target, they could calculate exactly where it would hit. Not so in the micro-world. The more carefully you measure some of these, the less certain the others become. You can't know all of them exactly at once; probabilities are all we will ever have.

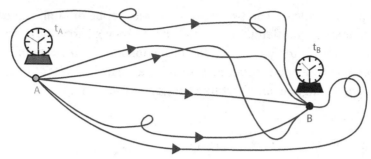

FIGURE 6.2 Particle path according to Feynman's sum over histories.

SOURCE: "The Sum over All Possibilities: The Path Integral Formulation of Quantum Theory," Einstein Online, http://www.einstein-online.info/spotlights/path_integrals.

Is there a grand design that unites the micro-world of probabilities with the world of experience, of certainty? There must be. As the philosopher Donald Davidson once said, "There is at most one world."[23] Hawking and his co-author turn to a formulation of Feynman to explain just exactly how the world of quantum physics becomes the world we experience. Feynman states that to determine the path of an elementary particle, we need to sum up all the paths a particle might take. Some are very probable, some very improbable. Figure 6.2 shows us some of these many paths from time t_A to time t_B.

It is the most probable paths of the particles of the micro-world that account for our everyday experience: in the diagram, the direct route from A to B:

> Feynman's theory gives an especially clear picture of how a Newtonian world picture can arise from quantum physics, which seems very different. According to Feynman's theory, the phases associated with each path depend upon Planck's constant [a very, very, very small number]. The theory dictates that because Planck's constant is so small, when you add the contribution from paths that are close to each other, the phases normally vary wildly, and so . . . they tend to add to zero. But the theory also shows that there are certain paths for which the phases [or probabilities] tend to line up, and so those paths are favored; that is, they make a larger contribution to the observed behavior of the particle. It turns out that for large objects, paths very similar to the path predicted by Newton's [laws] will have similar phases and add up to give by far the largest contribution to the sum, and so the only destination that has a probability effectively greater than zero is the destination predicted by Newtonian theory, and that destination has a probability that is very nearly one. Hence large objects move just as Newtonian theory predicts they will.[24]

Hawking and Mlodinow address a second cosmic mystery. While the contents of the universe obey the laws of physics, those laws do not in any way determine the constants on which they operate, constants like the speed of light and acceleration due to gravity; these seem entirely arbitrary. Even the laws that incorporate these constants seem arbitrary. Why should two masses attract each other directly according to their masses and inversely according to the distance between them? Hawking and his co-author show that while they are individually arbitrary, in the aggregate these laws and constants form a coherent whole, a grand design that does not seem in the least arbitrary.

We know this because if any of the fundamental constants, relationships, or laws had deviated significantly from their present status, we would not be here to discover and observe them. Our very existence depends on a series of accidents that may not be accidental. This opens the possibility that the characteristics of the cosmos form a coherent whole, a grand design. "There may well be," Hawking and his co-author say, "one unified theory that allows for the existence of structures as complicated as human beings who can investigate the laws of the universe."[25] We wouldn't be here if things were not pretty much as they are. From the beginning the universe was so ordered that we were bound to emerge: this is the strong anthropic principle.

The formation of carbon is a significant link in this almost miraculous chain of contingencies. Carbon is essential to our makeup, the foundational element of life on earth. If carbon had not formed, our emergence would not have been possible. But carbon didn't just happen; it required a universe full of stars. As these aged, they accumulated helium, two atoms of which unite to form beryllium, at this point an unstable element that almost immediately turned back to helium. As the stars aged further, however, their cores collapsed. As they collapsed, they got very hot. In this fiery environment, beryllium was transformed into carbon. But not enough carbon would have been manufactured by this process to take the next step toward life. This would remain the case until "the sum of energies of a beryllium nucleus and a helium nucleus [was] almost exactly the energy of a certain quantum state of the isotope of carbon formed, a situation called resonance, which greatly increases the rate of a nuclear reaction."[26] Even this carbon, however, was a prisoner of a dying star; it could take a next step toward the formation of life only when that star finally exploded so that its carbon was freed.

This almost miraculous fine-tuning includes the forces that hold carbon atoms together. If the strong force changed as little as 0.5%, if the electromagnetic force changed as little as 4%, then it is unclear that life as we know it could never have got started. Moreover,

> it turns out that it is not only the strengths of the strong nuclear force and the electromagnetic force that are made to order for our existence. Most of the fundamental constants in our theories appear fine-tuned in the sense that if they were altered by only modest amounts, the universe would be qualitatively different, and in many cases unsuitable for the development of life.[27]

In short, we reach the astonishing conclusion that if we "change [the] rules of our universe just a bit, . . . the conditions for our existence disappear!"[28] No wonder the great astrophysicist Fred Hoyle said that he did "not believe that any scientist who examined the evidence would fail to draw the inference that the laws of nuclear physics have been deliberately designed with regard to the consequences they produce inside the stars."[29]

Still, by applying Feynman's sum over histories to the universe, Hawking and Mlodinow argue away the implied need for a grand designer. In their view the quantum state of the universe can be expressed by analogy with Feynman's sum over histories, meaning there are unimaginably many universes besides ours (possibly an infinite number) just as there are an infinite number of paths from time t_A to time t_B in figure 6.2:

> The universe appeared spontaneously in every possible way. Most of these correspond to other universes. While some of those universes are similar to ours, most are very different. They aren't just different in details, such as whether Elvis really did die young or whether turnips are a dessert food, but rather they differ even in the apparent laws of nature. In fact, many universes exist with many different sets of physical laws.[30]

Since there is no evidence for a multiverse and very well may never be, we are in the realm of Greene's speculative sublime, the realm of scientific science fiction.

Hawking, the Man into Myth

"Active" may seem an odd adjective to apply to someone with Lou Gehrig's disease, but we must remember that the activity of theoretical physicists is largely mental: the myth of the lone theorist is no myth at all. While the CERN project that discovered the Higgs boson involved a staff of thousands, Peter Higgs, who had long before argued for its existence, worked alone. Newton worked alone; Faraday worked alone; the papers of 1905, Einstein's miraculous year, are the products of a single mind. Even today a team of theoretical physicists is not the norm. They collaborate, they talk to each other, but they predominantly work alone or at most with one or two others. The army of assistants, nurses, and technicians required for Hawking to function is unrelated to his creativity; these helpers concern only the maintenance of his ability to move from place to place and to communicate with others: to speak and to write things down. He does indeed resemble "the manager of a company,"[31] but the company that he manages is required solely because of his disability.

While it is absurd to think that anyone is better off as a consequence of a debilitating disease, it is far from absurd to think that Hawking has turned this disease to his advantage by controlling the social and intellectual network he has created out of necessity. Speaking of him, a colleague says:

> He doesn't look after children. He doesn't cook dinner. I'm late today because I went to check on my daughter at school. He doesn't have that. He sits in his chair and he has his computer. He can't talk much; that means he doesn't teach. He doesn't deal with the bureaucracy that university—all of us have to. He has 24 hours a day to contemplate. And if you are looking for a simple explanation for something, that's a good way to be. You don't want too much noise.[32]

Walt Woltosz, the man whose computer wizardry made it possible for Hawking to speak, makes the paradox clear—the independence of someone absolutely dependent:

> Now when Stephen wants his nurse—and his nurse is usually in her room in the back of the house—when he wants her to come where he is, he sends an environmental control signal that makes her door open. So that is his signal to her, and when she sees her door open, she knows that [Stephen needs her]. . . . He built a new home a few years ago, and his entire house is all set up with all these actuators. It is quite remarkable. . . . He is extremely independent, considering the level of physical disability he has.[33]

This dominance over his physical and human environment extends to the international society of theoretical physicists who routinely defer to his fame. When Hawking asks for a speaking slot at an international conference in Dublin, a slot is opened for him, even though it is well past the deadline for submissions.[34] This dominance extends even into the heart of theoretical physics. Because the equations that are central to physicists' creative process are virtually impossible for Hawking to manipulate, he has trained himself to internalize diagrams, especially a form of diagram invented by two close collaborators, Roger Penrose and Brandon Carter. In the words of physicist Alan Guth: "The cleverness of the Penrose diagram is that . . . it distorts space so that . . . all infinite space is mapped onto a finite-sized picture. The finite-size picture still shows all infinite space."[35] As a consequence, "a diagram can be equivalent to a hundred lines of algebra."[36] Through these diagrams, despite his disability, Hawking can exert his power over space and time; these diagrams enable Hawking's creativity as a theorist.

Those who dominate can become domineering. It is a trait all of Hawking's biographers ignore, in part because their contact with him is severely limited. Indeed, two of his biographers, co-authors John Gribbon and Michael White, never met with him at all.[37] For a depiction of Hawking at his worst we must turn to the memoir of his ex-wife, Jane Hawking:

It has to be said that there was a further dimension to traveling with Stephen, quite apart from the actual flying, which only increased my reluctance. Even with the help of the nurses, assistants and students, every such expedition was now indescribably nerve-racking, attended by some twenty to thirty pieces of luggage and so many unforeseen factors that an expedition to the Himalayas would have seemed like a children's picnic in comparison. The first nail-biting question was whether the taxi or limousine summoned at dawn for the drive to Heathrow would be large enough to take the luggage as well as the travelers. Check-in at the airport would be a confrontational experience. While we held up a long queue, the check-in clerk would find it difficult to suppress his or her disbelief and irritated frustration at the amount of luggage and the apparently impossible demands of the frail but command-ing body in the wheelchair. Whatever seating had already been allocated, Stephen's requirements admitted no opposition, and accordingly hasty rear-rangements would have to be made to seat him and his entourage wherever he chose. He would pay scant regard for the airline's normal policy for deal-ing with disabled people because he had devised his own. Come what may, he would insist on driving his wheelchair to the door of the plane, suspecting that to do otherwise might mean the loss *en route* of a vital part. Only there would he allow it to be dismantled with strict instructions as to the disposi-tion of the various parts: the main frame could be taken to the hold but all the bits and pieces, the armrests, footrests, headrests and cushions had to be stowed away in the cabin, in the overhead lockers or such spaces as the cabin crew could conjure up. He himself would then have to be carried on board. As often as not, the seat he had chosen would not be suitable after all and he would want to change yet again. The laptop computer would then have to be set up so that he could argue out the details of his gluten-free meal—if indeed one had been provided—with the steward. . . . Each of these demands, taken singly, was reasonable enough, even justifiable; all together they would fray the nerves of even the most rugged travelling companion.[38]

Taken singly, as Jane Hawking says, these demands are legitimate; after all, Hawking cannot function at all if an essential part of his electric wheelchair has gone missing. But that these are demands is unsettling. It is this that accounts for Jane Hawking's embarrassment and anger, and for his secretary's resentment: "He is very difficult, He is very demanding, very egotistical."[39] Both these women have second thoughts. The secretary stops mid-sentence and asks the interviewer to shut the recorder off. Jane Hawking revises her memoir, retitling it *Travelling to Infinity* and omitting the offending passage in its entirety. For a time, at least, they share with others what they do not dare share with Stephen Hawking.

This conspiracy of silence is generated by the force of Hawking's personality, his charisma, a characteristic "by virtue of which he is set apart from other men and treated as endowed with supernatural, superhuman, or at least specifically exceptional powers or qualities."[40] There is an aura about him, as if an oracle were speaking with the authority of a god. It is an aura his disability successfully enables:

Someone will ask him a question, and it takes a long time for him to construct an answer, and the answer comes, and it has to be simplified, because he can't otherwise express it, and so it comes back rather enigmatic. And so it's hard for young students, I think, particularly, because they see these enigmatic answers . . . and they have somehow to make this work, and it's difficult because they don't quite understand. . . . And it means he has a kind of oracle role, which is a strange one in a scientific community I think.[41]

To be an oracle is to be without emotions, dreams, desires, without a personality, without an ordinary life. It is to be only an oracle. Errol Morris, the director of the film of *A Brief History Time*, says of Hawking: "You never know if Stephen is annoyed by your question and thinks you're an idiot, or is it a joke? . . . It's a hall of mirrors."[42] His student Christophe Galfard opines: "I like to think we've got very friendly. I think he rather likes me in fact. And you know, you never really know, he could just hate everything I stand for."[43] The archive of Stephen Hawking being assembled at Cambridge will not contain a personal diary. As far as the world knows, as far as the world will ever know, Stephen Hawking has no more inner life than the machines on which he relies.

It is this oracular role that another film, *The Hawking Paradox*, incorporates. That Hawking actively participated in the film and controlled its image of him allows us to infer that he is happy with his oracular self; indeed, it seems to be the only Hawking Hawking wishes us to know. In the film he is described as a "dominant figure" in his field, a man who "astonishingly" turned his disability "to his advantage." He understands black holes "better than anyone else"; he experienced "one of the great insights of modern physics," an "elegant equation" that exhibits "his mathematical brilliance" and confirms "Hawking's reputation as a genius." When he proposes his radical idea that information permanently disappears from the universe, it creates "stunning confusion" because "his mathematics seemed air tight." Even when "the tide was turning against him," Hawking, "the most stubborn individual in the entire universe," resisted: "One man begged to differ." According to the director, William Hicklin, "The film was . . . about the process of how Hawking works on his ideas."[44] This certainly seems to be the impression Hawking wishes to give.

The secularization of society over the last two centuries in the West may not mean that religious feeling has diminished; it may mean only that it has shifted from traditional objects of worship to political leaders, news anchors, pop singers, sports stars, and at least one star in theoretical physics. Vladimir Lenin's charisma is exemplary in its effect: "I have seen Lenin speak to his followers. A small, busy, thick-set man under blinding lights, greeted by applause like thunder. I turned around and their faces were shining, like men who looked on God. Lenin was like that, whether you think he was a damnable Antichrist or a once-in-a-thousand-years' prophet. That is a matter of opinion, but when five thousand faces can light up and shine at the sight of him, as they did, and I saw it, then I say he was no ordinary individual."[45] According to a man who eventually became his staunch political enemy, Lenin had a "positively hypnotic effect upon people."[46]

Hawking's effect is analogous. Visiting from America, physics professor Karen Walton manages to procure a ticket to a Hawking lecture set to begin at eight in the evening:

> That chain of events led me to the Cockcroft Lecture Theatre by 6:45 p.m. on October 30, 1992, because I could not restrain my exuberance at the thought of being in the presence of the most famous scientist since Albert Einstein.

When I arrived seventy-five minutes before the lecture, I was not the first in line. Before me was an undergraduate student, and my arrival was followed by that of a gentleman in his sixties. My generation lies between the two.[47]

After the lecture, Hawking is asked about his belief in God. He answers that that would mean another hour's lecture. Walton comments: "From the roar of laughter and applause, it was clear that the audience was more than happy to stay."[48] After another lecture, this time before a professional audience, Hawking is asked what his next project might be. After a long pause, he answers: "I don't know." Hélène Mialet, Hawking's ethnographer, comments that the remaining audience "laughed and clapped—desperate to interpret the most slight of utterances as sign of profound wit."[49] In a footnote, she mentions that sociologist Harry Collins, a member of the audience, compared the event with a religious ceremony.[50] Hawking does not have audiences, he has worshipers; he does not have readers, he has fans.

A comparison with Newton, one of Hawking's heroes, is illuminating. Here was another man who surrounded himself with a coterie he dominated. Here was a man who after his death was worshipped as a secular saint, a man whose rooms at Trinity were transformed into a shrine. Privileged visitors were invited to these rooms, where "every relic of [his] studies and experiments were respectfully preserved to the minutest particular, and pointed out to me by the good old Vice-Master with the most circumstantial precision."[51] Alexander Pope's epitaph epitomizes this excess of adulation:

Nature and Nature's Laws lay hid in Night:
God said, "Let Newton be!" and all was light.[52]

In a parallel fit of excess, a famous sculptor, Louis-François Roubiliac, was commissioned to carve a full-size statue of Newton. The cost was £3,000 (over $600,000 in 2017 dollars).[53] Of this statue, Wordsworth wrote:

I could behold
The antechapel where the statue stood
Of Newton with his prism and silent face,
The marble index of a mind forever
Voyaging through strange seas of thought, alone.[54]

A statue of Hawking has also been commissioned; an archive of Hawking manuscripts and artifacts is being assembled in a "collective process of memorializing his identity."[55]

Conclusion

Chartres Cathedral is an hour away from Paris. You can take the train there, spend an hour, then return. On the other hand, you can spend the day, guidebook in hand, studying the architecture, the sculptures, the stained glass as the sun in its transit lights each window in turn. Like Chartres, Hawking's books have something to say to the casual visitor as well as to the avid student. Every traversal, no matter how cursory, will tell the reader something interesting about science and about a particularly interesting man in close touch with its sublime, anxious to share his experience with his readers.

The tale Hawking tells, a tale in which his professional biography and his science are inextricably intertwined, need not defeat the understanding of the general reader. We must remember that there is more than one way to read with understanding. Of course, readers are puzzled and amazed by what has been said about black holes, Hawking radiation, the no-boundary universe, time travel, and the grand design; they are expected to be puzzled and amazed, as puzzled and amazed as Lucy Hawking's George at this dazzling display of the mathematical and dynamic sublime, related to us by Stephen Hawking, the oracle who divines "the mind of God,"[56] a physicist who is an exemplar of the scientific sublime.

PART II | The Biologists

CHAPTER 7 | Rachel Carson
        | *The Ethical Sublime*

Rappaccini! Rappaccini! and is *this* the upshot of your experiment!
—NATHANIEL HAWTHORNE, "Rappaccini's Daughter"

RACHEL CARSON HAS become Saint Rachel, canonized time and again by the environmental movement. May 27, 2007, marked the 100th anniversary of her birth. In that year, the Cape Cod Museum of Natural History in Brewster, Massachusetts, hosted a major Rachel Carson centennial exhibition. The show was a partnership project of the museum and the US Fish and Wildlife Service, and it featured artifacts, writings, photographs, and artwork from Carson's life and career.[1] In 2012, the 50th anniversary of the publication of *Silent Spring* was commemorated by a Coastal Maine Botanical Gardens event and exhibit. From September 7 through October 23, the exhibit presented artwork, photos, and interpretive panels in the visitor center.[2]

Canonization, and the posthumous fame it bestows, comes at a price: the disappearance of the Rachel Carson whose work was driven by two forces. The first was the love of nature. A perceptive review of *The Sea Around Us* compares Carson with great science writers who share with her a love of nature:

It is not an accident of history that Gilbert White and Charles Darwin described flora and fauna with genius, nor that the great mariners and voyagers in distant lands can re-create their experiences as part of our own. They wrote as they saw and their honest, questing eye, their care for detail is raised to the power of art by a deep-felt love of nature, and respect for all things that live and move and have their being.[3]

The second force was the love of a woman, Dorothy Freeman, a person who in Carson's view made her later life endurable and her later work possible:

> All I am certain of is this: that it is quite necessary for me to know that there is someone who is deeply devoted to me as a person, and who also has the capacity and the depth of understanding to share, vicariously, the sometimes crushing burden of creative effort, recognizing the heartache, the great weariness of mind and body, the occasional black despair it may involve— someone who cherishes me and what I am trying to create, as well.[4]

It was this relationship that was paramount. Concerned about her legacy, she wrote to Dorothy that their time together contained not only their mutual love but "everything I shall be remembered for."[5]

Carson enjoyed the celebrity her success as an author bestowed, a celebrity that turned to fame after the publication of *Silent Spring*. More than seventy newspapers wrote editorials on the book.[6] Later in life, she received awards from the Garden Club of America, the American Geographical Society, and the National Audubon Society, along with election to the American Academy of Arts and Letters.[7] She also enjoyed adulation from afar in the form of fan letters and invitations to speak. Although she refused many prestigious speaking engagements, she accepted one closer to home. She spoke at Maine's Wiscasset High School on behalf of the Lincoln County Cultural and Historical Association.[8]

Despite the fact that home was more important than fame and that public engagements made her nervous, the spotlight could not always be avoided. In the wake of her television appearance on *CBS Reports*, "The Silent Spring of Rachel Carson," she received an invitation to testify before the Senate Subcommittee on Government Operations. She demurred, citing problems with her health. The invitation was immediately superseded by telephone call from Senator Abraham Ribicoff urging her to testify.[9]

Always she worried deeply that her fame might degrade her relationship with Dorothy:

> I am so very happy to feel that we *are* sharing these experiences just in the way I had hoped we might. You know, I'm sure, that what I feared was that in some way—through seeing me publicly cast in the role of "famous author"—that horrid barrier might rise again.[10]

Her books, "the everything I shall be remembered for," are a progress from the love of nature to its defense, from environmental rhapsody to environmental ethics. *Under the Sea Wind* testifies to her passion for nature. The book that follows, *The Sea Around Us*, represents an advance in skill and depth. It is her celebration of the vastness and power of the world's oceans, her evocation of the mathematical and dynamic sublime on a global scale. In *Edge of the Sea*, her next, the guidebook is reborn as literature, the unveiling of the lives of the creatures of the Atlantic shore, even the tiniest, as a source of wonder. *Edge* is testimony that the Kantian sublime exists, even in the remotest corner of the biosphere. *Silent Spring*, her last book, is a gripping apocalypse in modern dress, a polemic fueled by an implacable environmental ethics. It unleashes an unrelenting tide of indignation against those who would interfere with nature and, in their arrogance, destroy the very basis of the sense of wonder we were all born to experience, natural wonder seen through the lens of science.

Apprenticeship

Under the Sea Wind testifies more to Carson's love of nature than to her literary skill. Her model is the great nature writer Henry Williamson. Just as Williamson follows the fortunes of Tarka the otter and Salar the salmon, in *Under the Sea Wind*, Carson follows those of Scomber the mackerel and Anguilla the eel. A great admirer of Williamson, Carson nevertheless lacks his skill as a novelist, his ability to make his animals and people live indelibly in our imaginations.

There is a problem with literary representations of animal life, one that philosopher Thomas Nagel pinpoints. He objects to the employment of the imagination to enter another subjectivity, such as a bat's. Imagination "tells me only what it would be like for *me* to behave as a bat behaves. But that is not the question. I want to know what it is like for a *bat* to be a bat."[11] To do so would be to identify with animals, to become one with them, a seemingly impossible task.

This ineffability is a challenge to a writer who wants to limn fellow creatures from their own point of view rather than his own. But Williamson, a man who labored hard to cut out "untruths and inaccuracies, facile descriptions, inventions not of true imagination,"[12] meets this challenge with incomparable literary skill: he achieves literary sublimity by seeming to be at one with the creatures he describes. In his work, self seems annihilated: Williamson seems wholly transported into the creatures that star in his books. In *Salar*, some otters snare an unwary salmon:

> Gleisdyn had not struck the small otter deliberately. His strength was going
> from him in shock after shock of fear as he fled from the sight or smell of

one or another of the otters. Twice more he went down to the bridge, and upstream again, followed by a wide wave. On the third rush down river he came face to face with the bitch [otter], whose teeth as he slashed round bit into the wrist of his tail. His momentum tore him free, and in wildest fear he swam up the run again, driven by the feeling that he must get away from the ford. Past the ford he arrowed, and soon was threshing amidst flat stones and thin water, desperately driving himself forward, breaking his skin on sharper edges of larger stones. The otters ran upon him, curled their hair-spiky necks over him, clasping the cold body with their stumpy forelegs, and slicing off scales and skin with their teeth. Curving and slapping as he escaped, Gleisdyn writhed to the water, but the dog [otter] caught him by one of the fore fins and dragged him back, rising up and falling upon him to tear flesh from the thick part of the shoulder. Soon the large bones of the back were exposed, and the fish bled to death. When it ceased to move, they left it, for they were not hungry: this was their sport.[13]

Carson so admired this passage that she imitated it, or tried to:

When the island and its reef had been left a mile behind, the mackerel school was thrown into sudden panic by the appearance among them of a herd of some half-dozen porpoises which had risen to the surface to blow. The porpoises had been feeding on an underlying sandy shoal, where they were rooting out the launce who had buried themselves there. When the porpoises found themselves among schooling mackerel, they slashed at the little fish with their narrow grinning jaws, killing a few mackerel in sport. But when the school fled in swift alarm through the sea, they did not follow, for they had already gorged on the launce to the point of sluggishness.[14]

Both these accounts are fictions. Neither author is interested in what happened, the historical truth, but in what happens, truth to nature. Yet the distance between these two narratives is wide: his diamonds are her paste. Hers is a picture not out of life but out of books. Researching *Tarka*, Williamson "took to following the local hunt, the Cheriton Otter Hounds, who hunted along the local rivers, the Taw and Torridge and their tributaries, whose waters meet in the estuary beyond Barnstaple. HW also spent some time at a zoo . . . where he could watch otters at close range, noting particularly how much they actually played with water."[15] What Williamson experienced, Carson researched: "Carson did her homework, sometimes with the aid of a secretary, sometimes not. File folders were filled with correspondence, scientific studies, and reports."[16] Averse to sea voyages, Carson could not have observed the encounter she described; she could only employ her knowledge of the habits of porpoises and mackerel to create a plausible inkhorn account. "No, there is no such salmon as Gleisdyn," Williamson seems to say." No, I didn't actually see what I described. But it is truth nonetheless, derived from my experience. I describe only what I actually saw, at one time or another, as I wandered not far from my Devon cottage."

Williamson is not only more vivid than Carson, he is more accurate. Like any good scientist, he notices every relevant detail. Like any great literary artist, like Proust is his cork-lined room, he searches for the right word, and does not stop until he finds it: arrowed. Taking perceptual differences into consideration— factors such as the salmon's lateral line, a sort of radar—Williamson creates the conditions that enable us to identify with characters who just happen to be animals. We do not simply view but experience nature. Williamson's artistry places us inside the heads of salmon and otters: we experience their fear, we feel their pain. Unlike Carson, Williamson accomplishes—more accurately, he seems to accomplish— the impossible, a goal that is for master novelists their quotidian achievement. In *Portrait of the Artist as a Young Man* or *War and Peace*, in any masterly novel, we experience an imperceptible slide into persistent illusion, a movement from what it is like *for us* to be Stephen or Natasha to what it is like simply *to be* Stephen or Natasha; we experience a whole construction that fills every corner of the imagination. We experience in Joyce or Tolstoy the literary sublime; in Williamson we experience, conjointly, the literary and natural sublime.

Williamson presents us not only with real otters and salmon but with real fishermen and their antagonists, real fishery police, the water bailiffs. The former are sly natives speaking in Devon dialect, the latter pompous officials speaking an alien English:

> "For the last time I ask you, will you turn out that bag?" shouted the water bailiff. "Or shall I give you in charge?"
>
> "Aw, don't 'ee vex yourself so," said the skipper, in a gentle voice. "Here's an Easter egg for 'ee," and he dropped the heavy weight, and tugged the sack from the blubbery mass [of battered dolphin].
>
> "It's yours, Nosey Parker," yelled the fishwife as she staggered away, holding the arm of her husband and laughing stridently.[17]

In contrast, in *Under the Sea Wind*, in the very midst of a mackerel hunt, one commercial fisherman pauses to reflect:

He leaned over the gunwales, peering down into the dark water, watching the glow fade, seeing in imagination what he could not see in fact—the race and rush and downward whirl of thousands of mackerel. He suddenly wished he could be down there, a hundred feet down, on the lead line of the net. What a splendid sight to see those fish streaking by at top speed in a blaze of meteoric flashes![18]

Carson's fishermen as unrealistic as her fish. She describes rather than recreates an imagined experience; worse, it is *her* imagined experience.

A lack of novelistic skill also behind the major structural defect of *Under the Sea Wind*. It lacks narrative unity, the underlying source of the power of the two books Carson so admires. Salar and Tarka are the centers of consciousness of their respective books, just as Stephen is the center in *Portrait* or Gregor in *Metamorphosis*. Salar is a hero on his last quest: the final return from the sea to his spawning ground and death; like us, Tarka struggles in life and against death, even in his last, losing battle. *Salar* and *Tarka* have plots, a feature *Under the Sea Wind* singularly lacks. Its only unity is the progress of the seasons. There is no sense of narrative inevitability.

In *Under the Sea Wind*, Carson puts her own thoughts about the natural sublime into the mind of an anonymous fisherman, at sea only two years:

Not long enough to forget, if he ever would, the wonder, the unshakeable curiosity he had brought to his job—curiosity in what lay under the surface. He sometimes thought about fish as he looked at them on deck or being iced down in the hold. What had the eyes of the mackerel seen? Things he'd never see; places he'd never go.[19]

How in her next book did Carson give these thoughts life? How did she mobilize her considerable talents to produce not a pale imitation of Williamson but an enduring masterpiece of nature writing, a lasting evocation of the natural sublime seen through the lens of science?

Reinventing Nature Writing

The date of the first edition of Rachel Carson's first masterpiece, *The Sea Around Us* (1950) dates the science the book contains. The revised edition, published eleven years later, shares the same fate. Carson may reflect the one-time scientific consensus when she tells us that the Pacific basin once housed the moon, that the continents never altered their original positions, and that ontology recapitulates phylogeny. She may reflect the one-time historical consensus when she tells us that N. B. Palmer, the captain of the *Hero*, was the first to discover the continent of Antarctica[20] or that "the *Historia Norwegiae* records that Ingolf and Hjorleif found Iceland 'by probing the waves with the lead.' "[21] In all these cases she is mistaken.

None of this prevents *The Sea Around Us* from being a masterpiece, from successfully evoking the mathematical and dynamic sublime, the vastness and the power of the world's oceans seen through the lens of science. Her book is an ode to the sea, a panorama that fuses narrative and description into an integrated whole. The book begins with the formation of the earth and its moon, told at the pace of a geological clock. As time goes on, the clock ticks more rapidly, now tracking not geological but evolutionary time. Life begins in the sea and moves to the land, mammals emerge; eventually, human beings arrive. We discover that the "world is a water world, a planet dominated by its covering mantle of ocean, in which the continents are but transient intrusions of land above the surface of the all-encircling sea."[22]

With this phrase, "the surface of the all-encircling sea," Carson's organizational principle turns from time to space: "For the sea as a whole, the alteration of day and night, the passage of the seasons, the procession of the years, are lost in its vastness, obliterated in its own changeless eternity. But the surface waters are different. The face of the sea is always changing."[23] In the chapter "Mother Sea," we explore the complex geography of the ocean floor; the drift of sediments downward; the occasional emergence of islands. In "The Restless Sea" we investigate the changing relationships between sea and land; waves and tsunamis; currents; tides. In "Man and the Sea about Him," we deal with the relationship of the sea to our climate; we also deal with our attempts to extract wealth from the sea, especially its petroleum. In the final chapter, we return to the temporal order: the history of sea exploration and the developing art of navigation.

In "The Wind and the Water," Carson makes vivid the book's central themes: the forces of nature, our progress in understanding and accommodating to them, and—most important—our inability to control them, the shattering of the illusion that we are in charge. We hear from Lord Bryce and Charles Darwin; from Dumont d'Urville and Robert Cushman Murphy; from a naval observer and from an eyewitness to a tsunami. Henwood, an inspector of a mine that extends beneath the shoreline, creates a frightening picture of the improbable din just a few feet above, a matter-of-fact hymn to forces beyond our control, a portrait of the dynamic sublime:

> When standing beneath the base of the cliff, and in that part of the mine where but nine feet stood between us and the ocean, the heavy roll of the larger boulders, the ceaseless grinding of the pebbles, the fierce thundering of the billows, with the crackling and boiling as they rebounded, place a tempest in its most appalling form too vividly before me ever to be forgotten. More than once doubting the protection of our rocky shield, we retreated in affright; and it was only after repeated trials that we had confidence to pursue our investigations.[24]

Because Henwood's fear places him in imagination amidst the rampaging waters above, we are there too. Safe at home, we experience his fear; we experience the frisson of the sublime. How many dry-as-dust mine reports did Carson read until she stumbled on this paragraph of accidentally masterly prose, worthy of Longinus's attention.

That Carson's own prose is also masterly is no accident. When her subject requires it, she is the mistress of the plain style, moving from the structure of waves to the art of navigation, taking advantage the insight that mathematics can help us understand the dynamic sublime:

> Before constructing an imaginary life history of a typical wave, we need to become familiar with some of its physical characteristics. A wave has height, from trough to crest. It has length, the distance from its crest to that of the following wave. The period of the wave refers to the time required for succeeding crests to pass a fixed point. None of these dimensions is static; all change, but bear definite relations to the wind, the depth of the water, and many other matters. Furthermore, the water that composes a wave does not advance with it across the sea; each water particle describes a circular or elliptical orbit with the passage of the wave form, but returns very nearly to its original position. And it is fortunate that this is so, for if the huge masses of water that comprise a wave actually moved across the sea, navigation would be impossible.[25]

In contrast, in imagining the action of the wave, in allowing us to experience the dynamic sublime, Carson paints a picture with the mastery of a Joseph Turner:

> Young waves, only recently *shaped* by the wind, have a steep, peaked shape, even well out at sea. From far out on the horizon you can see them *forming*

whitecaps as they come in; bits of foam are *spilling* down their fronts and *boiling* and *bubbling* over the *advancing* face, and the final *breaking* of the wave is a prolonged and deliberate process. But if a wave, on coming into the surf zone, rears high as though *gathering* all its strength for the final act of its life, if the crest forms all along its *advancing* front and then begins to *curl* forward, if the whole mass of water *plunges* suddenly with a *booming* roar into its trough—then you may take it that these waves are visitors from some very distant part of the ocean, that they have *traveled* long and far before their final dissolution at your feet.[26]

Word choice and sentence form work together to animate this passage. This is not the jargon-laden and passive prose of scientific discourse. It is rich in the verb forms I have italicized: shaped, forming, spilling, boiling, bubbling, advancing, breaking, gathering, advancing, curl, plunges, booming, traveled. In the final sentence, three consecutive "if" clauses imitate the rhythm of an incoming distant wave as it rises, crests, falls, and dissipates. The linked "then" clause ends with a scientific observation: the wave has "traveled long and far" from a "distant part of the ocean" before dissolving at our feet in its "final act of its life."

In the composition of artistic prose there are at least two forms of versatility. Both seek the right words in the right order. In his *Euphues*, John Lyly, a lapidary artist, gives us a persistent network of balanced sentences tricked out with alliteration and laden with antithesis:

When parents have more care to leave their children wealthy than wise and are more desirous to maintain the name than the nature of a gentleman, when they put gold into the hands of youth where they should put a rod under their girdle, when instead of awe they make them past grace, and leave them rich executors of goods and poor executors of godliness, then it is no marvel that the son, being left rich by his father's will, becomes reckless by his own will.[27]

Nothing could be further from Carson's practice; she is no stylist for the sake of style. Whether describing the wave or the action of the wave, she suits the sound to the sense. She is at one with the wave, and because she is, we are as well. In *The Sea Around Us*, style and subject matter routinely converge to reveal the natural and scientific sublime.

The Sea Around Us is as fresh today as it was the day it was published. Read at first, in part, for the science it contained, it will continue to be read for the scope of Carson's vision, a gaze that encompasses the forces of nature that keep our planet in flux. While Carson holds our attention chapter by chapter with description and anecdote, she never permits our immersion in detail to undermine a panoramic view that integrates all into a single, well-articulated whole, a hymn to the sublime in nature, seen through the lens of science.

In her next book, *The Edge of the Sea*, Carson faces a new problem. For the first time, she undertakes the task of transforming her own experience into a literary form, a guidebook designed not to identify the creatures of the shore but to understand them. In *The Edge of the Sea*, Carson will pay the closest attention to plants and animals that generally escape our notice, from tiny snapping shrimp to the large white crabs of the Bahamas. Together, they form a web of life, an interactive network ruled by evolution, an ever-changing, self-perpetuating community that, when scrutinized, evokes our awe and astonishment.

The Guidebook Born Again

The Edge of the Sea began as "Guide to Seashore Life on the Atlantic Coast," a title highly unlikely to become a *New York Times* bestseller.[28] It is distinctly odd that Carson would ever contemplate writing such a guide. She was a nature writer, not a naturalist; she did not want her readers to identify the creatures of the eastern shore but to identify with them, to wonder at their intricacy and at the complexity of their lives, a microcosm of the biosphere. Carson's struggle *not* to write a guidebook, to free herself from the constraints of a project so unsuited to her talents, is well told by her two best biographers, Linda Lear and William Souder; I will not repeat it here. Nevertheless, a question they investigate is also central to me. Did Carson weave her descriptions of the individual plants and animals of the eastern shore into a single tapestry evocative of awe and astonishment? Did she make the guidebook genre her own?

Carson begins *The Edge of the Sea* by introducing us to "The Marginal World" of the Atlantic coast and to the "Patterns of Shore Life" in the three very different locales. She devotes a lengthy chapter to each: the rocky coast of the northern states, the sandy beaches of the central states, and the coral islands and mangrove swamps of Florida. A final chapter follows, a hymn to "The Enduring Sea." By itself, this simple structure is insufficient to turn a series of isolated descriptions into an engrossing story of one woman's experience with the plant and animal communities of the Atlantic coast. Neither can it do justice to the transformative character of that experience, the sense of wonder that Carson hopes to convey to the reader.

To discover how *The Edge of the Sea* accomplishes this transformation, we must look further than structure. In a letter to Paul Brooks, her editor at Houghton Mifflin, Carson provides a clue to the puzzle. She communicates her sense that she has just learned how to write the book she should have been writing all along. At last she can turn a guidebook into a real book, something people will read rather than consult:

Somewhere along in the Florida Keys chapter I decided that I have been trying for a very long time to write the wrong kind of book, and in dealing with the corals and mangroves and all the rest I seemed at last to fall into the sort of treatment that is "right" for me in dealing with this sort of subject. I seem to be doing the same sort of thing for the sand and I feel I can for the marshes, not yet begun. It means that I shall have to rewrite quite a lot of the rocks chapter—but that is part of writing, I guess.[29]

It is not enough, Carson tells us, to describe the coral coast; we must understand how it came to be: "To understand the living present, and the promise of the future, it is necessary to remember the past."[30] We cannot, of course, actually remember this distant time: "remember" is just her way of referring to the reconstructions scientists routinely make. The "promise of the future" is also a promise only science can make; only science can tell us that the earth is in constant flux, that our sense that we walk a solid planet is an illusion fostered by the brevity of our stay:

The corals that now form the substance of the eastern Keys built their reef during [the] Sangamon inter-glacial period, probably only a few tens of thousands of years ago. Then the sea stood perhaps 100 feet higher than it does today, and covered all of the southern part of the Florida plateau. In the warm sea off the sloping southeastern edge of that plateau the corals began to grow, in water somewhat more than 100 feet deep.[31]

Within this paragraph a map is embedded, showing in figure 7.1 the coast of Florida, a picture of two geological eras, the one superimposed on the other. Fully integrated into its surrounding prose, physically and conceptually, the map is scientific in what it says, evocative in what it suggests.

Having set the scene, Carson proceeds to the interactions within and among the animal and plant communities of the coral shore: the edge of the sea and its reefs, its tidal areas, its inland shallows, and its encroaching mangrove swamps. Each of these zones is subject to the rise and fall of the tides and to the cycles of day and night. Carson depicts the rising tide

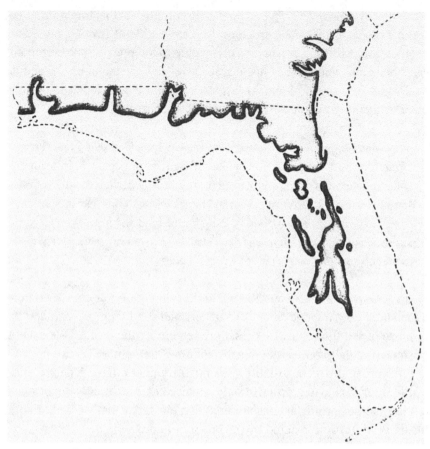

FIGURE 7.1 Florida past and present.

SOURCE: Rachel Carson, *The Edge of the Sea* (New York: New American Library, 1955), 194.

as a poignant convergence of the past and the present, allowing us to see its effect upon corals whose shape and extent are the work of thousands of years:

> The tides pouring in from the reefs and sweeping over the flats come to rest against the elevated coral rock of the shore. On some of the Keys the rock is smoothly weathered, with flattened surfaces and rounded contours, but on many others the erosive action of the sea has produced a rough and deeply pitted surface, reflecting the solvent action of centuries of waves and driven salt spray. It is almost like a stormy sea frozen into solidity, or as the surface of the moon might be. Little caves and solution holes extend above and below the line of the high tide. In such a place, I am always strongly aware of the old, dead reef beneath my feet, and of the corals whose patterns, now

crumbling and blurred, were once the delicately sculptured vessels that held the living creatures. All the builders now are dead—they have been dead for thousands of years—but that which they created remains, a part of the living present.[32]

It is especially under the cloak of darkness that the shore becomes alive:

> But that which seems quiescent—a dream world inhabited by creatures that move sluggishly or not at all—comes swiftly to life when day ends. When I have lingered on the reef flat until dusk fell, a strange new world, full of tensions and alarms, has replaced the peaceful languor of the day. For then hunter and hunted are abroad. The spiny lobster steals out from under the sheltering bulk of a big sponge and flashes away across the open water. The gray snapper and barracuda patrol the channels between the Keys and dart into the shallows in swift pursuit. Crabs emerge from hidden caverns; sea snails of varied shape and size creep out from under rocks. In sudden movements, swirling waters, and half-seen shadows that dart across my path as I wade shoreward, I sense the ancient drama of the strong against the weak.[33]

The passage begins with Carson on the edge of the sea, sharing her experience in the present tense: creatures *move*. When she tells us that she *has* lingered, however, we infer that her observations are the consequence of long-term intimate acquaintance. At this point, the scope of the passage broadens from human to evolutionary time: "I sense the ancient drama of the strong against the weak."

It is not Carson alone who observes the life of the shore; *The Edge of the Sea* is not a memoir. Scientific voices join hers, some in silent paraphrase, such as those embodied in the phrase "recent studies suggest";[34] some by name, such as "Dr. Ross Nigrelli and his associates";[35] and one, Alexander Agassiz, in his own voice. Others also join the chorus. Apuleius, the ancient author of *The Golden Calf*, makes an appearance; Columbus is given a voice.[36] There is another voice, however, far more important than these. An observer—it might be you—is startled by the sudden appearance of the very slim pipefish:

> The human swimmer drifting idly above the turtle grass—if he is patient enough and observant enough—may see something of other lives being lived above the coral sand, from which the thin flat blades of grass reach upward and sway to the motion of the water, leaning shoreward on a flooding tide and seaward on the ebb. If, for example, he looks very carefully he

may see what he had thought to be a blade of grass (so perfectly did it simulate one by form and color and movement) detach itself from the sand and go swimming through the water.[37]

Still, we must not miss the most important voice of all, Rachel Carson's. Carson consistently speaks in her own voice in the only book she wrote in that voice. It does not matter that "I" does not appear in a passage; it does not matter that at times she is relating the experience of others. These experiences are transformed into her own, imbued with her sense of wonder at their scientific significance. At last she achieves the center of consciousness Williamson achieved in *Tarka* and *Salar:*

> Here on these mangrove-fringed shores some of the pioneering mollusks and crustaceans are learning to live out of the sea from which they recently came. Among the mangroves and in the marshy areas where the tides rise over the roots of sea grasses there is a small snail whose race is moving landward. This is the coffee-bean shell, a small creature with a short, widely ovate shell tinted with the greens and browns of its environment. When the tide rises the snails clamber up on the mangrove roots or climb the stems of the grasses, deferring as long as possible the moment of contact with the sea. Among the crabs, too, land forms are evolving. The purple-clawed hermit inhabits the strip above the highest tidal flotsam, where land vegetation fringes the shore, but in the breeding season it moves down toward the sea. Then hundreds of them lurk under logs and bits of driftwood, waiting for the moment when the eggs, carried by the female under her body, shall be ready to hatch. At that time the crabs dash into the sea, liberating the young into the ancestral waters. Nearing the end of its evolutionary journey is the large white crab of the Bahamas and southern Florida. It is a land-dweller and an air-breather, and it seems to have cut its ties with the sea—all its ties, that is, but one. For in the spring the white crabs engage in a lemming-like march to the sea, entering it to release their young. In time the crabs of a new generation, having completed their embryonic life in the sea, emerge from the water and seek the land home of their parents.[38]

Can *pioneering* creatures really *learn* and *defer*? Can they really *cut their ties* with the sea? Can they really return *home*? Carson's is a subtle anthropomorphism that does not confuse us with our fellow creatures; rather, it helps us identify with them, linked as we are by extended chains of descent. "Recently" places us in the evolutionary flow; "is moving" turns space into a metaphor for time, an "evolutionary journey." It is these verbal networks, the products of Carson's artistry, that place us in the evolutionary tide in the same way that, at the beginning of this chapter, we were placed in an even wider context, the flow of geological time.

Our praise of Carson's prose must not mislead us into slighting the contribution of the book's illustrations to its impact. On nearly every page, Carson's words interact with Bob Hines's depictions, permitting us to experience the accommodations within communities at the edge of the sea fully, a synergy exemplified by the mutual accommodation of the loggerhead sponge and the snapping shrimp, a synergy of word and image illustrated in figure 7.2 and its accompanying prose:

There on the flats off some of the Keys, I have opened small loggerheads and heard the warning snapping of claws as the resident shrimp, small, amber-colored beings, hurried into the deep cavities. I had heard the same sound filling the air about me, as, on an evening low tide, I waded in to the shore. From all the exposed reef rock there were strange little knockings and hammerings, yet the sounds, to a maddening degree, were impossible to locate. Surely this nearby hammering came from this particular bit of rock, yet when I knelt to examine it closely there was silence; then from all around, from everywhere but this bit of rock at hand, all the elfin hammering was resumed. I could never find the little shrimps in the rocks, yet I knew they were related to those I had seen in the loggerhead sponges. Each has one immense hammer claw almost as long as the rest of its body. The movable finger of the claw bears a peg that fits into a socket in the rigid finger. Apparently the movable finger, when raised, is held in position by suction. To lower it, extra muscular force must be applied, and when the

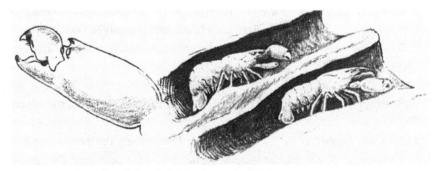

FIGURE 7.2 Snapping shrimp inhabit a loggerhead sponge.
SOURCE: Rachel Carson, *The Edge of the Sea* (New York: New American Library, 1955), 218.

suction is overcome, it snaps into place with an audible sound, at the same time ejecting a spurt of water from its socket.[39]

This is an easy match to the vividness Henry Williamson routinely achieves.

In two short centuries of industrial growth, we have disturbed, perhaps irreparably, the natural balance that preceded us, a balance at the core of *The Edge of the Sea*, a balance on which the long-term survival of our planet depends. We have consistently traded long-term consequences for short-term gains. It is this same loving concern for our earth and its creatures that motivates the polemic in *Silent Spring*, Carson's last book. *Silent Spring* is seriously misdescribed as a jeremiad against the pesticide industry; rather, it is a brief in favor of the balance of nature we so heedlessly undermine, the wonder we experience and that we hope our children and our children's children will also experience. It is nonsense to accuse of a lack of balance a work dedicated to the restoration of balance, to dismiss this balance as "so-called" or "obsolete."[40] As we turn to *Silent Spring*, we may say what Professor Pietro Baglioni said to Dr. Rappaccini: "Is *this* the upshot of your experiment!" For, in the interests of science, he has poisoned his own daughter.

The Apocalypse in Modern Dress

The book of Revelation presents us with an apocalyptic vision designed to strike terror in all but the saved:

> There was a tremendous earthquake, the sun turned dark like coarse black cloth, and the full moon was red as blood. The stars of the sky fell upon the earth, just as a fig tree sheds unripe figs when shaken in a gale. The sky vanished as though it were a scroll rolled up, and every mountain and island was jolted out of its place. . . . And when [the Lamb of God] had opened the seventh seal, there was silence in heaven about the space of half an hour. (6:12, 8:1)

After the final judgment, the author unveils vision of the future of the saved:

> I saw a new Heaven and a new earth, for the first Heaven and the first earth had disappeared, and the sea was no more. I saw the holy city, the new Jerusalem, descending from God out of Heaven, prepared as a bride dressed in beauty for her husband. (21:1)

In the first chapter of *Silent Spring*, "A Fable of Tomorrow," an apocalypse is imagined, one not precisely analogous, since no one will be among the saved:

A strange blight crept over the area and everything began to change. Some evil spell had settled on the community: mysterious maladies swept the flocks of chickens; the cattle and sheep sickened and died. Everywhere was a shadow of death. The farmers spoke of much illness among their families. In the town the doctors had become more and more puzzled by new kinds of sickness appearing among their patients. There had been several sudden and unexplained deaths, not only among adults but even among children, who would be stricken suddenly while at play and die within a few hours.[41]

The last chapter of *Silent Spring* presents us not with vision of the future, but with a choice:

We stand now where two roads diverge. But unlike the roads in Robert Frost's familiar poem, they are not equally fair. The road we have long been traveling has been deceptively easy, a smooth superhighway on which we progress with great speed, but at its end lies disaster. The other fork in the road—the one "less traveled by"—offers our last, our only chance to reach a destination that assures the preservation of our earth.[42]

The biblical vision has been secularized. Our heaven is the earth nature meant for us before the indiscriminate use of pesticides diverted us into incremental disaster. Our redeemer is an enlightened science, a family of disciplines sensitive to the intricacies of the web of life, the relationships that bind all in a single living and evolving community. This vision demands that our actions be animated by an environmental ethic. "A Fable of Tomorrow" has been called "a fabulous tale with a moral lesson,"[43] a lesson that applies to the book as a whole. Carson's ethical emphasis is clear, her sense that the enemies of nature lack "humility before the vast forces with which they tamper." She warns that turning the "terrible weapons" of science "against the insects it has also turned them against the earth."[44] She marshals her facts to turn belief into action. In Randy Harris's words, she wants us "to *know* that insecticides are dangerous. And to *do* something about it."[45]

Carson's vision of a web of life we must preserve is beautifully captured in Lois and Louis Darling's pen-and-ink drawings. In one, figure 7.3, a deer subsists on leaves while a predator waits in the wings, ready to

FIGURE 7.3 A deer feeds; a predator awaits its prey.

SOURCE: Rachel Carson, *Silent Spring* (Boston: Houghton Mifflin, 2002), 245.

subsist on the deer. To those sensitized to Carson's vision, the message is clear—in the absence of those leaves, the predator cannot survive. In the next illustration, figure 7.4, pesticide residue drips from leaves, the spraying apparatus X-ing out the natural world, the source of wonder. What is meant to preserve destroys.

Carson begins her polemic by detailing the dangers posed by the indiscriminate use of pesticides, an assault that disturbs the balance of nature. In marked contrast to the impassioned prose of her apocalyptic vision, in these passages she exercises restraint. The frisson of impending doom has been evoked; now its causes must be documented—the bare facts are themselves sufficiently persuasive. Of our war on our fellow creatures, Carson says:

> Professor Joseph Hickey and his students at the University of Wisconsin, after careful comparative studies of sprayed and unsprayed areas, reported the robin mortality to be at least 86 to 88 per cent. The Cranbrook Institute of Science at Bloomfield Hills, Michigan, in an effort to assess the extent of bird loss caused by the spraying of the elms, asked in 1956 that all birds thought to be victims of DDT poisoning be turned in to the institute for examination. The request had a response beyond all expectations. Within a

FIGURE 7.4 A pesticide gun and the tree it kills.
SOURCE: Rachel Carson, *Silent Spring* (Boston: Houghton Mifflin, 2002), 7.

few weeks the deep-freeze facilities of the institute were taxed to capacity, so that other specimens had to be refused.[46]

But it is not only robins that are in danger. When insecticide leads the way, homicide soon follows:

> The second major group of insecticides, the alkyl or organic phosphates, are among the most poisonous chemicals in that world. The chief and most obvious hazard attending their use is that of acute poisoning of people applying the sprays or accidentally coming in contact with drifting spray, with vegetation coated by it, or with a discarded container. In Florida, two children found an empty bag and used it to repair a swing. Shortly thereafter both of them died and three of their playmates became ill.[47]

At times, the violations of the balance of nature are so outrageous that Carson's indignation boils over. The effects of Diedrin is "notorious"; it strikes "with terrible effect at the nervous system" and causes "appalling destruction of wildlife."[48]

Indignation is justified anger. This means that Carson has a duty to rail at the conduct that stems from collective stupidity, from those who fail to realize that "for each of us, for the robin in Michigan or the salmon in the Miramichi, this is a problem of ecology, of interrelationships, of interdependence."[49] We are right to be angry when faced with indiscriminate pesticide use that interferes with the checks and balances that nature provides, the intricacies that allow communities of life to flourish. When these are ignored, disaster follows. To create more grassland for cattlemen, 10,000 acres of sage lands in the Bridger National Forest were sprayed with pesticide. The spraying killed the willows as well as the sage. With the death of the willows, the moose that fed on them disappeared. So did the beavers that turned these willows into dams. With the beavers gone, their dams deteriorated and the lake those dams created vanished. With the lake gone, its trout disappeared.[50] Indeed, indiscriminate pesticide use interferes with evolution itself. After spraying, only those pests survive who are immune to these poisons; only they reproduce, an effect is especially dangerous in the case of such insect-borne diseases as malaria. We also evolve, of course, but our defenses develop at a pace far too slow to matter.[51]

We might think that the pesticide industry would bear the brunt of Carson's indignation; we would be mistaken. True, early in the book she points to the rapid increase in pesticide production and the industry's enormous profits.[52] But then the industry virtually disappears from its pages. The brunt of Carson's indignation is reserved, instead, for municipal, state, and federal agencies, groups presumed to be dedicated to the public good and responsible to the people at large. While for the pesticide industry, it is natural to expect the profit motive to prevail, from government we have a right to expect responsible conduct in the public interest:

> The "agricultural engineers" speak blithely of "chemical plowing" in a world that is urged to beat its plowshares into spray guns. The town fathers of a thousand communities lend willing ears to the chemical salesman and the eager contractors who will rid the roadsides of "brush"—for a price. It is cheaper than mowing is the cry. So, perhaps, it appears in the neat rows of figures in the official books; but were the true costs entered. . . the wholesale broadcasting of chemicals would be seen to be more costly in dollars as well as infinitely damaging to the long-range health of the landscape and to all the varied interests that depend on it.[53]

In Revelation, God is the redeemer; in *Silent Spring*, enlightened science plays that role. It is a science that has given us two effective and environmentally sound weapons against insect and plant pests. The first is selective spraying. The mound-building habits of fire ants give sprayers specific targets, enabling them to destroy 90–95%.[54] One the other hand, the failure of ragweed eradication exemplifies the price of ignoring the laws that science reveals, the sources of natural control:

> Ragweed, the bane of hay fever sufferers, offers an interesting example of the way efforts to control nature sometimes boomerang. Many thousands of gallons of chemicals have been discharged along roadsides in the name of ragweed control. But the unfortunate truth is that blanket spraying is resulting in more ragweed, not less. Ragweed is an annual; its seedlings require open soil to become established each year. Our best protection against this plant is therefore the maintenance of dense shrubs, ferns, and other perennial vegetation.[55]

There is a problem with Carson's reliance on science, one she acknowledges only near the end of her book: "the mystifying fact that certain outstanding entomologists are among the leading advocates of chemical control."[56] There is nothing mystifying about this: scientific opinion about chemical controls is divided. Sometimes Carson's refusal to recognize that division tempts her to misrepresent the science on which her case depends. She mentions that the English Ministry of Agriculture declared a ban on arsenic-based insecticides; she fails to mention that the ban was voluntary.[57] She mentions that a British experimenter was sickened over a long period after the topical application of DDT; she fails to mention his conclusion that "the general consensus of opinion, based on experiments on animals and observations on man, is that D.D.T. used with discretion does not constitute a hazard to human health."[58] But Carson can also use divided scientific opinion to strengthen her case. In an early draft, scientific uncertainty sows doubts about the need for policy change. Because DDT can damage the liver, "we have some intimations as to what the ultimate consequences [of its use] may be, but it is too early to know the full story."[59] In *Silent Spring* the doubt expressed in this early draft is transformed into a call for action in the face of troubling uncertainty:

> All these facts—storage at even low levels, subsequent accumulation, and occurrence of liver damage at levels that may easily occur in normal diets, caused Food and Drug Administration scientists to declare as early as 1950

that it is "extremely likely the potential hazard of DDT had been underestimated." There has been no such parallel situation in medical history. No one yet knows what the ultimate consequences may be.[60]

In such passages, "her dramatization of possible risks invited her readers to manufacture political certainty in the face of scientific uncertainty."[61]

It is well to keep in mind that Carson does not advocate the banning of DDT. "It is not my contention," she says, "that chemical insecticides must never be used,"[62] a conclusion later research resoundingly supports.[63] Nevertheless, DDT was banned in the United States, largely as a consequence of her polemic, hardly a balanced view of the dangers of chemical pesticides. It was not banned, however, in the Third World, where it did yeoman service in curbing malaria. When DDT spraying ceased in Sri Lanka, malaria cases increased from nearly zero to 1.5 million in an epidemic that lasted from 1968 to 1970. DDT also led to a dramatic improvement in crop yields. In Asia, the Green Revolution, aided by pesticides, increased crop yields dramatically between 1965 and 1968. The greatest danger in this application was to farm workers, presumably because proper precautions were not taken.[64]

Conclusion

Rachel Carson took time to mature as an author, to discover her strengths. We would not want to judge Dvořák by his first two symphonies, Jackson Pollock by his first sketches, or James Joyce by *Stephen Hero*. *The Sea Around Us* was her first success. Playing to her strengths, she created a vision of the forces of nature, one that endures to this day. *The Edge of the Sea* also embodies this vision. In this book, Carson sees nature in miniature, the world in a grain of sand. To dismiss *The Edge of the Sea* on the grounds of its narrow scope would be like dismissing the lapidary perfection of Mozart's Adagio in B Minor because it is "not as good as" his *Jupiter* Symphony. *The Edge of the Sea* is also the book in which Carson developed her sense of the intricate interactions within and among communities of life, an insight central to *Silent Spring*, her enduring brief in favor of a future free from indiscriminate pesticide use.

Some of the science Carson relates in her books has been superseded; that is the way of science. What has not been superseded is her peerless evocation of wonder that the close scrutiny of the natural world can reveal, the natural sublime seen through the lens of science. It is her literary quality that makes her work endure, inspiring the environmental movement and teaching us all to respect and admire the natural world and to struggle against its enemies with the knowledge science provides. *Silent Spring* aims to change the way we behave, to change our future, person by person, industry by industry, government by government, society by society. Carson's rising tide of indignation at our misbehavior has its source in her perception that some environmental policies are making our future dystopian. Have we not in two short centuries irredeemably poisoned the air, earth, and water? It is Carson's optimism in the face of these unpalatable realities, an optimism tempered by realism, that has made *Silent Spring* the enduring classic it is, a sermon for her times and for ours. Like Revelation, *Silent Spring* gives us hope.

CHAPTER 8 | Stephen Jay Gould's Books
The Balanced Sublime

Of all things the measure is man, of the things that are, that they are,
and of things that are not, that they are not.

—PROTAGORAS

IN *QUESTIONING THE MILLENNIUM*, Stephen Jay Gould tells the story of an autistic child, a mentally challenged young man who nevertheless can tell in a flash the day you were born given the date you were born. Gould studies his subject until he understands the way this feat is accomplished. He realizes that every 28 years there is a consistent unit—28 extra days over the standard 52 weeks plus seven days for leap years, the equivalent of five weeks. His subject

> had added up extra days laboriously until he came to 28 years—the first span that always adds exactly the same total number of extra days, with the sum of extra days exactly divisible by seven. Every 28 years includes 35 extra days, and 35 extra days makes five weeks. You see, he *had* given me the right answer to my question—but I had not understood him at first. I had asked: "Is there anything special about the number 28 when you figure out the day of the week for dates in different years?" and he had answered: "Yes . . . five weeks."
>
> May we all make such excellent use of our special skills, whatever and however limited they may be, as we pursue the most noble of all our mental activities in trying to make sense of this wonderful world, and the small part we must play in the history of life. Actually, I didn't quote his beautiful answer fully. He said to me: "Yes, Daddy, five weeks." His name is Jesse. He is my firstborn son, and I am very proud of him.[1]

This anecdote exemplifies Stephen Jay Gould's passionate interest in science; equally, it testifies to his dispassionate pursuit of the science that interests him, a probing skepticism that with patient effort yields the correct answer. For Gould the scientist it is of no interest that his experimental subject is his son. Still, the motivation for his exploration would be lost if Jesse Gould were not Jesse *Gould*, an exemplary person not merely to be examined but understood, not merely understood but cared about, loved.

Jesse Gould's story is also exemplary because it exists at the intersection of three sets of leading ideas that are the subject of Gould's three most important popular science books. Jesse Gould's autism is the consequence of nature, not nurture, an explanation that rightly absolves his parents of guilt for his misfortune. But there is a downside to such explanations. They have been employed time and again to misattribute to nature what is in effect a dubious mathematical abstraction: general intelligence. This distortion is the subject of *The Mismeasure of Man*. When Jesse Gould devised an algorithm to fix the day of any date, he employed an aspect of time's cycle to make sense of time's arrow. *Time's Cycle, Time's Arrow: Myth and Metaphor in the Discovery of Geological Time* concerns the use of each of these ideas to construe and to misconstrue the history of the earth. Jesse Gould's autism is a contingent event, the result of a clash between two orders of nature, one leading generally to normal births, the other to occasional abnormalities. *Wonderful Life: The Burgess Shale and the Nature of History* tells us that were it not for the extinction of dinosaurs—another contingent event—we would have been denied the opportunity to evolve; we would have been bypassed by the gradualism that characterizes the Standard Theory of evolution. In all these instances, the scientific sublime is properly evoked only if the right balance is struck between two competing ideas; only then does our sense of wonder track the true state of nature in all of its complexity.

Mismeasuring Man

The Mismeasure of Man argues that a proper balance has not been struck in understanding intelligence when it is claimed that it is only measureable by a one-size-fits-all test and is determined by heredity. To Gould, if scientists support this claim, they are worse than mistaken; to act in this way is to employ science in a disreputable cause—to mismeasure man. To Philip Morrison, writing in *Scientific American*, Gould's book was "a persuasive chronicle of prejudice in science, founded on scrupulous examination of the record, enlivened by the talent of a gifted writer." To Morrison, "this volume takes on some of the sinister appeal of a tale of heinous crime. The epigraph bears the weight of the story. The young Darwin wrote: 'If the misery of our poor be caused not by the laws of nature, but by our institutions, great is our sin.' "[2]

To Gould, it is an abuse of science to employ intelligence testing to rank individuals and groups from smart to dumb, and then to assign the "dumb" permanently to an underclass. This abuse is just one form of bio-determinism, the conviction that intelligence is a comprehensive mental trait that a test can definitively measure. True, the proponents of bio-determinism seem to have science on their side. For example, proponents argue that studies have demonstrated that "black Africans" have an average IQ substantially lower than that of "Latino," "white," "Asian," and "Jewish" Americans.[3] Moreover, detractors think they have a sound Darwinian explanation as to just why Caucasians are intellectually superior:

> The farther north the populations migrated out of Africa, the more they encountered the cognitively demanding problems of gathering and storing food, gaining shelter, making clothes, and raising children during prolonged winters. Consequently, as the original African populations evolved into present-day Europeans and East Asians, they did so by moving in the direction of larger brains and greater intelligence.[4]

Intelligence tests measure verbal, mathematical, and abstract reasoning; test takers are given a single grade that purports to measure of innate intelligence, the factor g. Gould argues that intelligence is a concept far broader, that g is of little value as a measure of human potential. In support of this position he mentions the work of prominent psychologists like Louis Leon Thurstone, J. P. Guilford, and Howard Gardner, all of whom advocate theories of multiple intelligences. Gardner's theory, perhaps the most elaborate, divides intelligence into seven types.[5] For visual-spatial intelligence, think of an architect like Frank Gehry; for bodily-kinesthetic intelligence, think of a ballet star like Margot Fonteyn; for musical intelligence, think of a composer like Wolfgang Mozart; for interpersonal intelligence, think of a peace negotiator like George Mitchell; for intrapersonal intelligence think of a saint like Teresa of Avila; for linguistic intelligence, think of a novelist like Jane Austen; for logical-mathematical intelligence, think of a scientist like Alan Turing. The notion of general intelligence is not only too narrow; it also makes a hash of the lives of the justly famous.

Ulysses Grant wrote a masterly autobiography but couldn't tell one tune from another; he led Northern armies to victory but made a poor showing as president. Joe DiMaggio had a fifty-seven-game hitting streak but Marilynn Monroe divorced him, citing among other things his lack of intelligent dinner conversation.

It is simply bizarre to deny that Fonteyn dancing *Ondine*, Mitchell negotiating with Sinn Fein, or Gehry sketching the design of the Weisman Art Museum on a dinner napkin are not engaged in intelligent behavior. Gardner also has no use for the common practice of ranking children on the basis of IQ testing: He "would prefer to spend more resources helping learners understand and develop their individual intelligence profiles and spend less resources testing, ranking, and labeling them."[6] These criticisms do not mean that intelligence tests are not useful. Alfred Binet, who originated these tests, employed them not to rank students but to help single out those who needed special attention. Indeed, Gould found these tests useful in diagnosing his autistic son, Jesse.

The villain in Gould's narrative is the misuse of Charles Spearman's factor analysis to create a dubious artifact: general intelligence. Factor analysis is a mathematical technique that reduces the variability among observed variables. This useful tool is, however, open to abuse. Spearman, for example, asserted that variations in scores on an intelligence test could be reduced to a single unobserved variable, g, represented by a single number—IQ. He is explicit about the bio-determinism behind this point of view:

> The Function appears to become fully developed in children by about their ninth year, and possibly even much earlier. From this moment, there normally occurs no further change even into extreme old age. In adult life, there would seem no appreciable difference between the two sexes. The Function almost entirely controls the relative position of children at school (after making due allowance for difference of age), and is nine parts out of ten responsible for success in such a simple act as Discrimination of Pitch. Its relation to the intellectual activity does not appear to be of any loosely connected or auxiliary character (such as willingness to make an effort, readiness in adaptation to unfamiliar tests, or dexterity in the fashion of executing them), but rather to be intimately bound up in the very essence of the process.[7]

To make factor analysis comprehensible to general readers—a necessary move if they are to detect the flaw in Spearman's reasoning—Gould employs not mathematics but diagrams. Figure 8.1 shows how factor analysis reduces four test variables, two verbal and two mathematical, to one measure of central tendency, Spearman's *g*. It seems as if *g* is really there, summarizing for us how the brain actually works and why some brains are smarter than others. In figure 8.2 Gould shows that this is an illusion: the same mathematical technique that made *g* appear can make it disappear, simply by rotating *g* into a set of two primary mental abilities (PMAs).[8] In

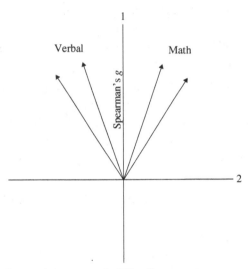

FIGURE 8.1 Plot of one primary mental ability, Spearman's *g*.
SOURCE: Stephen Jay Gould, *The Mismeasure of Man* (New York: Norton 1996), 284.

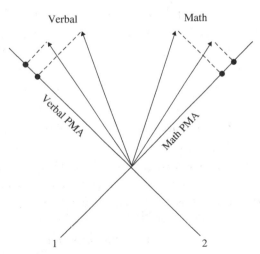

FIGURE 8.2 Two contrasting primary mental abilities.
SOURCE: Stephen Jay Gould, *The Mismeasure of Man* (New York: Norton 1996), 285.

this graph, the two mathematical and two verbal tests are associated with separate axes instead of a single axis.

Gould puts his finger on the fundamental problem of diehard bio-determinists: "the notion that such a nebulous, socially defined concept as intelligence might be identified as a 'thing' with a locus in the brain and a definite degree of heritability—and that it might be measured as a single number, thus permitting a unilinear ranking of people according to the amount they possess."[9] Reification is the source of this error, a process that isolates the result of a battery of tests from its proper context, manipulates it mathematically, and finally turns that result into an object called "intelligence." For Gould as for Gardiner there is something "fundamentally wrong with models which conceptualize intelligence in terms of a finite number of linear dimensions."[10] Today, factor analysts agree "that it is unlikely that a factor general to all abilities produces individual differences in all of what are regarded as indicators of human intelligence."[11]

Given his provocative thesis, it is unsurprising that *The Mismeasure of Man* received a hostile response in scientific quarters heavily invested in intelligence testing. Writing in *Nature*, Steve Blinkhorn called the book "a masterpiece of propaganda, researched in the service of a point of view rather than written from a fund of knowledge. For the best propaganda requires not the suppression or distortion of facts but their careful selection, emphasis and juxtaposition."[12] Blinkhorn seems to have a point. Was Gould's claim true that psychometricians seriously influenced the passage of the 1924 immigration act? No one could deny the racist intent of the act, which limited immigration to 2% of those already in the country according to the 1890 census—a benchmark chosen deliberately to limit the immigration of such undesirables as Jews and Italians. As a result of the act, immigration dropped from 242,000 in 1931 to 36,000 in 1932, of whom fewer than 3,000 were Jews;[13] Italian immigration dropped from 222,260 in 1921 to 6,203 in 1925.[14] While all this is true, Mark Snyderman and R. J. Herrnstein point out that Gould gets his history wrong.[15] He gives too much credit to the political influence of psychometricians; they had no direct influence whatever on laws restricting immigration. Perhaps Gould should have settled for a weaker claim: that these scientists had in common with the American Congress a persistent bias against Jews, southern Europeans, and African Americans, and that anyone familiar with their work would have been presented with a scientific justification for that bias.

A parallel weakening needs to be made for Gould's contention that the work of Cyril Burt played a major role in the formation of the notorious British "11+ exam," the battery of tests that decided the educational fate of children between 11 and 12, channeling them into higher education or, in the case of low scores, permanently into an underclass.[16] Burt's work was irrelevant. Franz Samelson was right to point out that 11+ "developed slowly out of the 'free place examination' for grammar school scholarships, instituted before any IQ tests existed. The testers' growing influence produced the eventual inclusion of an IQ test in and a new rationale for 11+, but not the examination itself or its social functions."[17]

Gould also wrongly argues that Samuel Morton's study of the cranial capacity of his large collection of skulls was influenced by racial bias, that Morton was a good scientist gone wrong because of a result of his commitment to bio-determinism. In separate articles, John Michael and Jason Lewis and his co-authors argue that although Morton was a racist, his racism did not significantly influence his science.[18] But when Lewis and his collaborators claim that there is no evidence that Morton linked cranial capacity to intelligence, they are clearly mistaken. Of the Shoshone Indian skulls Morton wrote:

Heads of such small capacity and ill-balanced proportions could only have belonged to savages; and it is interesting to observe such remarkable accordance between the cranial developments, and mental and moral faculties. Perhaps we could nowhere find humanity in a more debased form than among these very Shoshones, for they possess the vices without the redeeming qualities of the surrounding Indian tribes; and even their cruelty is not combined with courage. . . . A head that is defective in all its proportions must be almost inevitably associated with low and brutal propensities, and corresponding degradation of mind.[19]

Moreover, Lewis and his collaborators made serious errors in calculation. When these are corrected, "when one takes appropriately weighted grand means of Morton's data, and excludes obvious sources of bias including sexual dimorphism, then the average cranial capacity of the five racial groups in Morton's collection is very similar. This was probably the point that Gould cared most about."[20] Michael Weisberg's corrected table confirms this conclusion (table 8.1).

Morton also had no reason to believe that cranial capacity was linked to intelligence at all, any more than phrenologists like Franz Joseph Gall had a reason on their side when he claimed that bumps on the head were linked to mental capacities. True, we know today that a link between skull size and mental capacity does exist. Leigh Van Valen gives a correlation of 0.30 as a reasonable estimate,[21] and J. Philippe Rushton alleges that MRI studies place the figure at 0.40.[22] This means that cranial capacity accounts for between

TABLE 8.1 Comparative Cranial Capacities

MORTON'S RACIAL CATEGORY	MORTON'S REPORTED 1849 AVERAGE CRANIAL CAPACITIES (IN³)	RECONSTRUCTION OF THE 1849 CRANIAL CAPACITY TABLE (IN³)
Modern Caucasians	85	85
Mongolians	82	84
Malays	85	85
Americans	79	83
Africans	83	83

9% and 16% of intelligence. But these figures give no comfort to contemporary bio-determinists. Van Valen cautions us that this correlation does not mean causation and that, in any case, these figures are useless in dealing with individuals. Their real significance is evolutionary: cranial capacity expanded over time, but expansion terminated in the Neanderthal era.[23]

While Gould's history of mental testing requires some correction in retrospect, the essential point of *The Mismeasure of Man* has withstood all assaults; Philip Morrison's estimate of the value of his book has been proven correct. Gould's results cast a shadow over not only Morton's research program but also those of J. Philippe Rushton and Arthur Jensen, 20th-century scientists who insist that in this particular case correlation is indeed causation, that cranial capacity unerringly points in the direction of racial differences in intelligence. This continuing interest in cranial capacity and racial difference would seem to support Gould when he says in the second edition of *Mismeasure* that "the same bad arguments recur every few years with a predictable and depressing regularity;"[24] that scientific sublime is continually being hijacked in the interest of social injustice. The proper understanding of intelligence requires tipping the balance in favor of multiple intelligences over a general intelligence that is "rankable, genetically based, and minimally alterable."[25]

Misconstruing Geology

Gould begins *Time's Arrow, Time's Cycle* by quoting one his scientific heroes on the value of geology:

> Such views of the immensity of past time, like those unfolded by the Newtonian philosophy in regard to space, were too vast to awaken ideas of sublimity unmixed with a painful sense of our incapacity to conceive a plan

of such infinite extent. Worlds are seen beyond worlds immeasurably distant from each other, and beyond them all innumerable other systems are faintly traced on the confines of the visible universe.[26]

The author of this passage, Charles Lyell, is a central figure in Gould's book, as is James Hutton. Both would be featured in any serious discussion of the history of geology. The third, Thomas Burnet, would not. Although Gould places the good bishop's *Sacred Theory of the Earth* (written in the 1680s) right beside Hutton's *Theory of the Earth* (1780s) and Lyell's *Principles of Geology* (1830s), he insists that he has not made a category mistake. These three works are alike in that they see the earth through the lens of one of two central ideas: time's arrow or time's cycle.

At first glance, Burnet's *Sacred Theory* seems like no more than an idiosyncratic version of the biblical story. We are witness to a sequence of cosmic, geological, and human events that begins in Eden, carries through to the Noachian deluge, and ends in the final fiery conflagration that precedes the thousand-year reign of Christ. But *Sacred Theory*'s innovation— Gould's raison d'être for placing this book beside those of Hutton and Lyell—is Burnet's belief that the history of the earth must proceed without the benefit of divine intervention. Burnet's is a deistic vision in which God sets the universe in motion once and for all; indeed, it is a sign of God's perfection that He need not interfere. The Noachian flood is a unique natural event, though foretold by Moses, speaking with the voice of God. As a natural event, the flood cannot have been local. It must have been worldwide; otherwise humankind could have fled to higher ground. Moreover, precipitation cannot account for the flood: forty days of rain cannot have produced so much water. At the time of the flood, therefore, a crack in the earth must have opened and water from the earth's interior must have poured out, an event recorded by Moses when he said that "the fountains of the great abyss were broken open" (Gen. 7:11). At that time,

> the force of the vapors increased, and the walls weakened which should have kept them in, when the appointed time was come that an all-wise Providence had designed for the punishment of a sinful world. The whole fabric broke, and the frame of the earth was torn in pieces, as by an earthquake; and those great portions or fragments into which it was divided fell down into the abyss, some in one posture, some in another.[27]

You may doubt that any construal could turn these speculations into science. But Gould observes that Burnet's insistence on the exclusion of the miraculous makes him "more committed to the reign of natural law" than Sir Isaac Newton.[28] In a long letter to Burnet, Newton conjectured that each of the six days of creation was extended well beyond twenty-four hours, insisting that "where natural causes are at hand God uses them as instruments in his works, but I do not think them alone sufficient for the creation and therefore may be allowed to suppose that amongst other things God gave the earth its motion by such degrees and at such times as was most suitable to the creatures."[29]

Although Burnet cannot possibly be seen as legitimate precursor to contemporary geology, Gould is right to make what for him is the central point: Burnet sees the flood as natural, unique, and unrepeatable. He is correct that Burnet's insistence on the historicity of geology—on time's arrow—sets him apart from his otherwise more distinguished successors, Hutton and Lyell. For Burnet insists that geology consists only of such unrepeatable events: "Whatsoever molders or is washed away from [the mountains] is carried down into the lower grounds and into the sea, and nothing is ever brought back again by any circulation. Their losses are not repaired, nor any proportionable recruits made from any other parts of nature."[30]

In contrast, Hutton and Lyell insist that "geological history" is an oxymoron. Deluded by the brevity of our lives, we mistake time's cycle for its arrow; we miss through this misconstrual the slow and steady circuit of destruction and repair that surrounds us everywhere. Hutton was the first to argue for this view. As Gould notes, it is he who first insisted that geological phenomena provide "palpable proof that the earth does not decline but once into ruin; instead . . . time cycles the products of erosion in a series that shows, in Hutton's most famous words, 'no vestige of a beginning—no prospect of an end.' "[31] Although Lyell follows in his footsteps, unlike Hutton, he is forced to defend this view against stronger and stronger countervailing arguments. Through the long publishing history of *Principles of Geology* (12 editions over 45 years), Lyell insists on the prevalence of time's cycle. Every edition of this extremely popular book[32] contained this same paean to a steady state:

> There can be no doubt that periods of disturbance and repose have followed each other in succession in every region of the globe; but it may be equally true, that the energy of the subterranean movements has been always uniform as regards the *whole earth*. The force of earthquakes may for a cycle of years have been invariably confined, as it is now, to large but determinate spaces, and may then have gradually shifted its position, so that another

region, which had for ages been at rest, became in its turn the grand theater of action.[33]

For this view, evolution constitutes a problem, a theory entirely underpinned by time's arrow. In his first edition, in 1830, Lyell asserts that the unique timeline of evolution may be an illusion, that in the future "the huge iguanodon might reappear in the woods, and the ichthyosaur in the sea, while the pterodactyls might flit again through the umbrageous groves of tree-ferns";[34] in the last edition this same endorsement remains.[35] Of course, Lyell must agree that the appearance of humans counts as an exception to any steady-state theory. "Exception," however, is the key word. Lyell asserts that our ascendency over our fellow creatures is moral and spiritual, not physical. Our uniqueness is not part of the natural world:

> Our study and contemplation of the earth, and the laws which govern its animate productions, ought no more to be considered in the light of a disturbance or deviation from the system, than the discovery of the satellites of Jupiter should be regarded as a physical event affecting those heavenly bodies. Their influence in advancing the progress of science among men, and in aiding navigation and commerce, was accompanied by no reciprocal action of the human mind upon the economy of nature in those distant planets; and so the earth might be conceived to have become, at a certain period, a place of moral discipline and intellectual improvement to man, without the slightest derangement of a previously existing order of change in its animate and inanimate productions.[36]

Geology is a historical science, one in which the arrow of time, a central idea favored by Burnet, plays the major role. America and Europe will never again unite; the Grand Canyon will never become shallower. But time's cycle, the central idea advocated by Hutton and Lyell, is also important in understanding the history of the earth. Ice ages come and go; as they do, the sea level rises and falls. The poles change their positions; the poles reverse. Neither time's arrow nor time's cycles will generate the authentic geological sublime; a balance is required between these two central ideas.

Getting Evolution Wrong

The title of Gould's *Wonderful Life: The Burgess Shale and the Nature of History* refers both to the much loved Frank Capra film and to the wonder evoked by the fossils discovered in an obscure Canadian quarry near Banff. Gould's *Wonderful Life* is a hymn to the "new view of life that [these fossils] have inspired."[37] For him the site itself evokes sublimity: "One of the most majestic settings I have ever visited."[38] More important, so do the eighty thousand fossils collected there. Here is Gould at his literary best, comparing these fossils with sacred objects:

> The animals of the Burgess Shale are holy objects—in the unconventional sense that this word conveys in some cultures. We do not place them on pedestals and worship from afar. We climb mountains and dynamite hillsides to find them. We quarry them, split them, carve them, draw them, and dissect them, struggling to wrest their secrets. We vilify and curse them for their damnable intransigence. They are grubby little creatures of a sea floor 530 million years old, but we greet them with awe because they are the Old Ones, and they are trying to tell us something.[39]

The book and the film it is named for share a theme. George Bailey's guardian angel teaches him a lesson: he replays the tape of life as if George had never been born, a fact with tragic consequences. Contingency also looms large in the case of the Burgess Shale. If those animals had been an evolutionary highway, rather than a dead-end, we would not be around to study them. If the gradualism at the core of the Modern Synthesis were all evolution was about, we never would have evolved.

In *Wonderful Life*, Gould places Harry Whittington and Simon Conway Morris at the center of the action, bringing to life the contingency of the scientific innovation that brought to life the contingency of evolutionary history:

> Whittington is meticulous and conservative, a man who follows the paleontological straight and narrow, eschewing speculation and sticking to the rocks—exactly the opposite of anyone's image for an agent of intellectual transformation. Conway Morris, before the inevitable mellowing of ontogeny, was a fiery Young Turk, a social radical of the 1970s. He is, by temperament, a man of ideas, but happily possessed by the patience and *Sitzfleisch* [Yiddish for backside] needed to stare at blobs of rock for hours on end. In legend, the Burgess reinterpretation would have emerged as a tense synergism between these men—Harry instructing, pleading caution, forcing attention to the rocks; Simon exhorting, pushing for intellectual freedom, nudging his reluctant old mentor toward a new light. . . . I don't think that any of this occurred, at least not overtly.[40]

In recounting the bumpy ride to scientific discovery, Whittington and Conway Morris are Gould's protagonists; his antagonist is Charles Wolcott, head of the Smithsonian Institution, a leading 20th-century paleontologist. Gould summarizes Wolcott's scientific virtues: "a great geologist, an indefatigable worker, a noted synthesizer, a central source of power in the social hierarchy of American science." He also mentions his chief limitation: he was "not, fundamentally, an intellectual innovator."[41] Between 1909 and 1917, Wolcott led expeditions to the Burgess Shale, eventually gathering some 80,000 fossils. But he went awry in placing these in existing animal groups, characterizing them as missing pieces of standard evolutionary history, fated to undergo gradual change over time: "In short, he shoehorned every last Burgess animal into a modern group, viewing the fauna collected as a set of primitive or ancestral versions of later, improved forms."

Five decades later, Whittington, Conway Morris, and Derek Briggs came to realize that many of these fossils were unique, a "wrong" turn in evolutionary history that led to a dead end. The extinction of these ancient animals was contingent. In Gould's words, "Wind back the tape of life to the early days of the Burgess Shale; let it play again from an identical starting point, and the chance becomes vanishingly small that anything like human intelligence would grace the replay."[42] This does not mean that evolution is the science of higgledy-piggledy. Evolutionary theory can certainly make predictions. It can foresee that the beaks of Darwin's finches will vary with the relative abundance and character of the seeds on which the birds feed. But evolutionary theory cannot predict that finches will migrate to the Galapagos or indeed that there will be finches at all. Accordingly, Gould defies any paleontologist to predict that of the Burgess animals *Aysheaia* will survive and *Marrella* will not. Because evolution is also a historical phenomenon, it is deeply contingent, beyond the explanatory scope of any theory:

> A historical explanation does not rest on direct deductions from laws of nature, but on an unpredictable sequence of antecedent states, where any major change in any step of the sequence would have altered the final result. This final result is therefore dependent, or contingent, upon everything that came before—the unerasable and determining signature of history.[43]

The Burgess fossils support this claim. They rested in the Smithsonian storage rooms virtually undisturbed from the second decade of the 20th century until the seventh. When Whittington, Conway Morris, and Briggs re-examined them, they discovered that the earlier identifications of Wolcott required serious correction. *Wonderful Life* is the story of that correction, a process in which these compressed animal remains buried for hundreds of millions of years were meticulously dissected and in which their scattered body parts were reunited into single and unique organisms. Whittington, Conway Morris, and Briggs worked as taxonomists, not theorists; their main task was reconstruction and identification. Their identification, however, was a contingent event that was grist for Gould's theoretical mill.

At the center of Gould's narrative are not only his scientists but also the creatures they studied, in particular, "some twenty to thirty kinds of arthropods that cannot be placed in any modern group."[44] Of these, because they fit into no existing animal phylum, five "weird wonders" stand out. They are described in language typical of the natural sublime: "Whereas *Marrella* and *Leanchoilia* may be beautiful and surprising, *Opabinia*, *Wiwaxia*, *and Anomalocaris* are awesome—deeply disturbing and thrilling at the same time."[45]

How did Walcott's identifications go wrong? Gould attributes his errors to theoretical biases in favor of "direct predictability and subsumption under invariant laws of nature."[46] To support these biases, Walcott misidentified body parts and even doctored his photographs to accord with his biases. For example, having misidentified *Aysheaia pedunculata*, Walcott added "an imaginary head."[47] This was not a one-time deviation. In a monograph on *Marrella*, Whittington writes of Walcott's photographs, "several are heavily retouched to the point of falsification of certain features, notably the representation of the supposed mandible, maxilla, and maxillula."[48] Walcott's most startling error is a double misidentification. First, he misconstrues one animal part as the abdomen of a new species, *Anomalocaris gigantean*. In reality, it is a feeding appendage of an animal far larger than any in the Burgess find, an animal nearly four inches long. A second misidentification follows. Walcott names and identifies *Peytoia nathorsti* as a jellyfish, failing to notice that no jellyfish lacks tentacles and that none has an enormous hole in its center, misidentified as a mouth. Properly identified, both of Wolcott's finds are parts of a single animal, *Anomalacaris*. Gould lets us see the problem by juxtaposing figures 8.3 and 8.4.

Although Walcott cannot be absolved of careless work, the task of the correct decipherment of the fossils of the Burgess Shale clearly also exceeded his visual-spatial capacity, a form of intelligence that Whittington, Conway Morris, and Briggs have to an extraordinary degree.

FIGURE 8.3 *Peytoia nathorsti.* Xs mark four large lobes.

SOURCE: Charles D. Walcott, *Cambrian Geology and Paleontology*, vol. 2, *Middle Cambrian Holothurians and Medusae* (Washington, DC: Smithsonian Miscellaneous Collection, 1911), 56.

It is visual-spatial intelligence that enables its possessors to perceive and to re-create a three-dimensional visual world from "a squashed and horribly distorted mess."[49] The reproduction of a long-dead animal via its fossil "involves a sensitivity to the various lines of force that enter into a visual or spatial display . . . feelings of tension, balance, and composition."[50] It has been said that Nikola Tesla's mental images of his innovative machines "were more vivid than any blueprint. [His] inner imagery was sufficiently acute that he could build his complex inventions without drawings. Further, he claimed to be able to test his devices in his mind's eye."[51] Whittington, Conway Morris, and Briggs are his heirs.

The case of the fifteen extant specimens of *Aysheaia* is an example of their skill. The fossils were created when *Aysheaia* "were caught up by a turbulent cloud of sediment in suspension which was moving down-slope, and the bodies were buried as the current slowed down and the suspension settled out."[52] Over time, the fossils were compressed by the pressure of overlying rock. When discovered, they consisted of their part, the animal itself, and their counterpart, the impression of the animal in the rock above it. For Walcott, who mistakenly viewed the fossils as two-dimensional

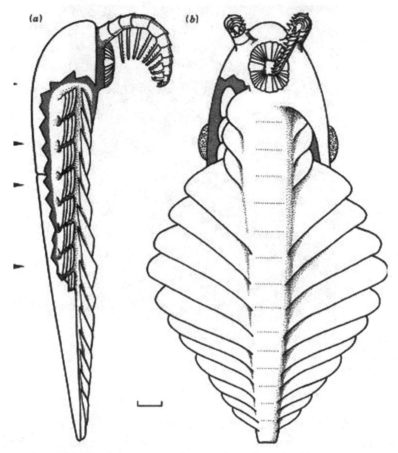

FIGURE 8.4 Whittington and Briggs's "largest Cambrian animal," *Anomalocaris.*

SOURCE: H. B. Whittington and D. E. G. Briggs, "The Largest Cambrian Animal, Anomalocaris, Burgess Shale, British Columbia," *Philosophical Transactions of the Royal Society B* 309, no. 1141 (1985): 569–609. For a recent morphological revision, see Allison C. Daley and Gregory D. Edgecombe, "Morphology of *Anomalocaris canadensis* from the Burgess Shale," *Journal of Paleontology* 88, no. 1 (2014): 68–91.

imprints, these counterparts were of no interest. But Whittington's fundamental insight—that the fossils were preserved in three dimensions, not two—meant that part and counterpart were both important in any proper reconstruction. In remarkable feats of digital dexterity, infinite patience, and visual-spatial intelligence, these three-dimensional fossils were dissected and their inner parts revealed. In Whittington's words, "I think [this procedure] was vital. Of course it took hours and hours, but you saw everything yourself, and various things could sink in gradually. . . . It is so exciting to find hidden things. It is an incomparable thrill to reveal the hidden structure in the rock."[53] Whittington is relating his sense of the scientific sublime, experienced as the miraculous revelation of the structure and function of a tiny creature from the remote past.

Figure 8.5 models the effects of the compression whose expansion is Whittington's task. At the top, we see two limb pairs as they were at burial; below are the same limbs under various degrees of compression. The last two frames illustrate the part-counterpart relationship. In (*d*) the lower half is the part; in (*h*) the upper half is the part. L is the left limb, R the right; sp is the large spine on the limb.

For Gould the meticulous reconstructions of Whittington, Conway Morris, and Briggs call for a major shift in the balance between two key evolutionary ideas: the predictability that gradualism permits and the contingency for which gradualism cannot account. In figure 8.6 Gould creates

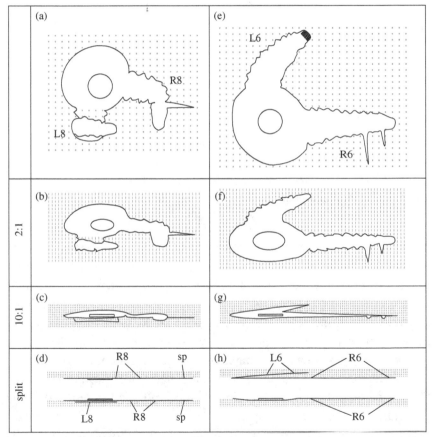

FIGURE 8.5 Fossil compression ratios and fossil preservation in part and counterpart.

SOURCE: H. B. Whittington, "The Lobopod Animal *Aysheaia pedunculata* Walcott, Middle Cambrian Burgess Shale, British Columbia," *Philosophical Transactions of the Royal Society London B* 284 (1984): 172.

The Cone of Increasing Diversity

Decimation and Diversification

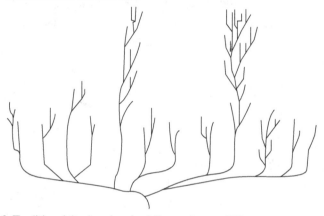

FIGURE 8.6 Traditional (*top*) and revised (*bottom*) tree of life.

SOURCE: Stephen Jay Gould, *Wonderful Life: The Burgess Shale and the Nature of History* (New York: W. W. Norton, 1989), 46.

a simple illustration of the difference between these two visions of evolutionary development. The top image represents Walcott's cone of increasing diversity, a view that dictated that the Burgess Shale fossils "had to be classified as primitive forms within modern groups, or as ancestral animals that might, with increased complexity, progress to some familiar form of the modern seas."[54] The bottom image "presents a new vision reflecting the lessons of Burgess Shale. The maximum range of anatomical possibilities arises with the first rush of diversification. Later history is a tale of restriction, as most of these early experiments succumb and life settles down to generating endless variants upon a few surviving models."[55] In Gould's view, contingency and restriction rule after an initial explosion when life proliferated "like a bacterial cell alone in an agar plate."[56] Gould's position remains controversial. As evidenced by reassessments of the Burgess Shale evidence by Simon Conway[57] and Briggs,[58] it could be out of balance itself.

Conclusion

Stephen Jay Gould was a gladiator at heart. At 31, with Niles Eldridge, he published his first foray into controversy—the theory punctuated equilibrium, a form of contingency that would seriously modify prevailing notions of all-pervasive gradualism, the core of the modern synthesis.[59] According to Gould and Eldridge, the well-known and well-publicized gaps in the fossil record would never be filled by fossils just waiting to be found; instead, they were a clear indication that at most times nothing evolutionary happens: stasis prevails. Significant events like speciation occur only on occasion. When they do, they occur rapidly—rapidly, that is, when seen against the vast eons of the geological time scale.

There was a second foray into the limelight. At 38, Gould co-authored "The Spandrels of San Marco," an attack on the adaptationist program, the view that natural selection works continually to optimize the adaptive traits of animals and plants.[60] To Gould and his co-author, Richard Lewontin, the fundamental body plans of plants and animals functioned as serious constraints on their adaptive possibilities, a fact the American and British theorists tended to ignore. Gould and Lewontin also deplored their tendency to tell just-so stories: first, the Eskimo face is engineered to deal with the cold; second, the Eskimo face is designed to maximize the ability to chew, two tales spun an evidential vacuum.

In both of his controversial articles, Gould's target was an overemployed leading idea, a misguided commitment that, he felt, shaped the evolutionary research program to its detriment, the idea that every change is gradual, the idea that every trait is adaptive. The books that are our focus in this chapter make the same point in areas as disparate as psychometry, geology, and evolutionary history. In each, Gould attacks a leading idea that he feels has shaped a scientific research program to its detriment: biodeterminism in *The Mismeasure of Man*, time's cycle in *Time's Arrow, Time's Cycle*, and in *Wonderful Life* the strictly orderly and gradual expansion of the evolutionary tree. His is not an attack on leading ideas themselves as guides to scientific research programs; it is a plea for balance, a plea that the sense of wonder the study of science evokes persistently motivate the search for a truth more complex than previously imagined.

CHAPTER 9 | Stephen Jay Gould's Essays
Experiencing the Sublime

But these have never trod
Twice the familiar track,
Never, never turned back
Into the memoried day,
All is new and near
In the unchanging Here
Of the fifth day of God,
That shall remain the same,
Never shall pass away.
On the sixth day we came.

—EDWIN MUIR, "The Animals"

"THIS VIEW OF LIFE," Stephen Jay Gould's long-running essay series, forms a massive refutation of any charge that in popularized science truth and entertainment are incompatible. These essays invariably "follow two unbreakable rules," Gould writes: "I never lie and I strive mightily not to bore you."[1] In the prologue to the collection *Bully for Brontosaurus*, Gould is more explicit concerning his first rule: "No compromises with conceptual richness; no bypassing of ambiguity or ignorance; removal of jargon, of course, but no dumbing down of ideas (any conceptual complexity can be conveyed in English)."[2] Concerning the second rule, Gould is silent: he practices but does not reflect on his ability to shock us, to imitate in prose the last line of Muir's "The Animals." On the surface, a statement of biblical fact, its position at the poem's end belies its factual status. We experience instead the shock of our existential state, burdened by history, by memory, by the apprehension of death. Muir has sprung a surprise, giving us an experience Longinus long ago described, the literary sublime. By literary means as well, in his essays on science and its history, Gould can spring analogous surprises. In Gould, however, the literary sublime is always in the service of its scientific counterpart.

I single out a group among his three hundred essays, each of which is so structured that we vicariously experience the process of discovery. Their twists and turns seem at first to lead nowhere; no picture emerges until, suddenly, one does, a surprise that evokes the experience of the scientific sublime. A very young Adam Smith—an Adam Smith well before *The Wealth of Nations*—tells us how us how surprise opens a path to this sublime: "When one accustomed object appears after another, which it does not usually follow, it first excites, by its unexpectedness, the sentiment properly called surprise, and afterwards, by the singularity of the succession, or order of its appearance, the sentiment properly called wonder."[3]

Experiencing Scientific Discovery

"The Panda's Thumb of Technology" is a set of variations on a theme, twists and turns whose relationship to each other, and to science, is initially puzzling. Although each perplexes by its seeming irrelevance, each is eventually integrated into a whole, a surprising unification that reveals a hitherto hidden link between biological and cultural evolution. Given the title, we are bound to be puzzled by the essay's introductory paragraph, a retelling of a well-known story from the Bible. As thanks for his victory over his foes, Jephthah agrees to sacrifice the first living thing he sees on his return; to his dismay, it is his daughter. This story reminds Gould of Handel's oratorio of the same name. Its libretto gives the tale a positive twist: Jephthah's daughter does not die. To emphasize the significance of this outcome, the librettist quotes from Alexander Pope's *Essay on Man*: "Whatever is, is right."

For Gould, Pope's epigraph encapsulates our most persistent biological delusion: that the world of living things is designed. In fact, all living things are the result of natural forces, of selection operating on random genetic variation. Contingency rules. Constrained by structures that have accumulated over time. Pandas are a notorious example. Over evolutionary eons, they have lost their thumbs; these have become permanently part of the panda's paw. If pandas were to grasp the bamboo that became their favorite food, a pseudo-thumb had to be derived from another bone. Nature is not an architect working from a plan but a weekend handyman working within existing structures, constraints on what is possible.

A quarter of his essay has passed and Gould still hasn't reached the topic his title had advertised. Or has he? We have learned that the tale of Jephthah is relevant to evolutionary theory. His is a story of how the past constrains the present in ways over which we have little control; it presents us with results that are—to say the least—far from optimal. To think otherwise, to think like Pope, is to exhibit, in Gould's view, "the bias of 'adaptationism'—the notion that everything must fit, must have a purpose, and in the strongest version, must be for the best."[4] The Panda's thumb is evidence to the contrary. Nature is invested not in the good but in the good enough.

But is there a panda's thumb of technology? Is there a legitimate analogy between biological and cultural evolution? Are both, like Jephthah, subject to past constraints that work against optimal solutions? The analogy between biological and cultural is, Gould rightly points out, fraught with pitfalls: there are so many *dis*analogies. Living beings evolved over millions of years through such mechanisms as natural selection, sexual selection, and genetic drift. As they evolved, they diverged: we are only one distant twig at the far end of the tree of life. In contrast, the form of cultural evolution we call technological change is relatively swift, purposeful, and convergent. After Röntgen stumbled on X-rays, little time was wasted before they were integrated into medical and industrial practice.

But this is not always the case. With persistence of the QWERTY keyboard, a keyboard that slows typists to accommodate now defunct mechanisms, we have a demonstration that biological evolution and its cultural counterpart do have something in common:

> My main point . . . is not that typewriters are like biological evolution (for such an argument would fall right into the nonsense of false analogy) but that both keyboards and the panda's thumb, as products of history, must be subject to some regularities governing the nature of temporal connections. As scientists, we must believe that general principles underlie structurally related systems that proceed by different overt rules. The proper unity lies not in false applications of these overt rules (like natural selection) to alien domains (like technological change), but in seeking the more general rules of structure and change themselves.[5]

In this paragraph, the fulcrum of his essay, Gould shifts from the deliberate intersection of cultural contexts—biblical, musical, poetic—to a single, overriding context, that of science itself. The voice of authority emerges from that of the genial entertainer: the conclusion is foreshadowed that technology and evolution have contingency and prior constraint in common. The panda's thumb and QWERTY have in common "the more general rules of structure and change." From this truth, Gould draws a surprising final inference, an evocation of wonder at the scope of contingency:

> But why fret over lost optimality. History always works this way. If Montcalm had won the battle on the Plains of Abraham, perhaps I would be typing *en français*. If a portion of the African jungles had not dried to savannas, I might still be an ape in a tree. If some comets had not struck the earth (if they did) some 60 million years ago, dinosaurs might still rule the land, and mammals would be rat-sized creatures scurrying about in the dark corners of the world. If *Pikaia*, the only chordate [precursor of the vertebrate] of the Burgess Shale, had not survived the great sorting out of body plans after the Cambrian explosion, mammals might not exist at all. If multicellular creatures had never evolved after five-sixths of life's history had yielded nothing more complicated than an algal mat, the sun might explode a billion years hence with no multicellular witness to the earth's destruction.[6]

In this essay, while Gould is thinking his way to a theoretical conclusion, we are right there thinking beside him.

"If Kings Can Be Hermits, Then We Are All Monkeys' Uncles" takes a different path to an analogous conclusion: it moves from the unabashedly personal to the scientifically significant, giving Gould another way to make us virtual discoverers. He begins with a personal anecdote:

> We learn from our errors, most of all from our shameful mistakes. I therefore begin with a story at my own expense. Many years ago, one of my students told me about her father's brother, a severely retarded man of childlike disposition. When she described him as "my uncle," I did a mental double take (and fortunately said nothing, so the shame of my error remained internal until now). I said to myself, "Uncles are wise people who render free advice (not always worthwhile) and take you to baseball games; how can a person with such limits be an uncle?" I then kicked myself (also metaphorically) and continued the soliloquy: "He is her father's brother; he is

therefore her uncle pure and simple; *uncle* is a genealogical term of relationship, not a functional concept of action; he is as good and as true an uncle as any man who ever lived."[7]

The jocular tone, the anecdote and its emotional resonance: these are signs of the personal essay—of Montaigne alive and kicking. But first impressions can be misleading; they are deliberately so in an essay whose topic is how misleading first impressions are. To do science is to learn systematically to discard first impressions, and this Gould—and we—proceed to do.

With an average weight of ten pounds and an average leg spread of three feet, king crabs are one of the impressive inhabitants of the deep. Not so the various species of hermit crab, creatures an inch or two long who make their homes in abandoned snail shells. It is not just size: the two look very different. King crabs look like crabs; hermit crabs look like tiny lobsters. Despite these first impressions, however, scientists have long believed that king crabs "are closest cousins of hermit crabs."[8] This is because they noticed some similarities in adult anatomy, suggestive but not compelling evidence of kinship. More compelling were similarities in their respective larvae or embryos. Even more compelling was that "'crablike' form is an oft-repeated trend in decapod [ten-legged] crustaceans."[9] None of this evidence was decisive. It was DNA that settled the matter. But words are not enough to make Gould's case. Figure 9.1 is essential to our understanding.

In the next paragraph, we are the tourists, he, the tour guide:

We now come to the truly remarkable point. Notice the two lower species in the upper clump—the king crab (*Paralithodes camtschatica*) and its close relative *Lithodes aequispina.* Now consider the species in the two major subclumps of this larger group formed at the separation of the right and left-twisting lines at division A. The upper subclump represents species of the genus *Pagurus,* the standard hermit crab of any textbook or local seashore. But now look at the lower subclump—and note that it includes two further species of the genus *Pagurus* along with the two species from the king crab line. In other words, king crabs are so close to hermit crabs by the proper criterion of propinquity that they actually branch off *from within* a narrowly restricted genealogical grouping so conventional in form and behavior that all species have been included in the canonical genus of *Pagurus!*[10]

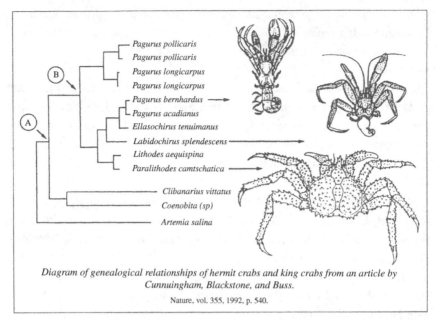

Diagram of genealogical relationships of hermit crabs and king crabs from an article by Cunnuingham, Blackstone, and Buss.

Nature, vol. 355, 1992, p. 540.

FIGURE 9.1 Crab genealogy.

SOURCE: Stephen Jay Gould, *Dinosaur in a Haystack* (New York: Harmony, 1995), 396.

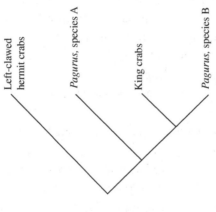

Chart by Mark Abraham.

FIGURE 9.2 Crab lineage.

SOURCE: Stephen Jay Gould, *Dinosaur in a Haystack* (New York: Harmony, 1995), 398.

In the three images in this figure we can see a transformation in the direction of "carcinization," or crablikeness; we can also see just when left- and right-twisting crabs split off (Point A) and when right-twisting crabs did (Point B).

Figure 9.2, a simpler diagram, makes these evolutionary relationships even clearer.

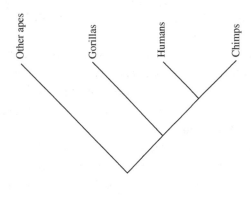

Chart by Mark Abraham.

FIGURE 9.3 Human lineage.

SOURCE: Stephen Jay Gould, *Dinosaur in a Haystack* (New York: Harmony, 1995), 398.

While this genealogical proximity is surprising, the effect is somewhat mitigated by its limited appeal. Most of us do not care a great deal about relationships among crabs. It is at this point that Gould springs his surprise. In figure 9.3, he retains the form of his diagram but alters its labels.

So we are really monkey's uncles. With Gould as our tour guide, we have made our way from personal anecdote to evolutionary theory. In retrospect, Gould's title, however jocular, serves the same purpose—exactly the same purpose—as scientific titles in professional journals. They summarize the content of the article; they are, as it were, an abstract of the abstract; in capsule form, they are a peephole into the scientific sublime.

Experiencing the Limits of Scientific Theory

While there is no scientific method, no one way to do science, there are scientific methods, a constantly increasing number that lead, with luck, to the discovery of significant patterns and laws. All methods are alike, however, in one respect: they involve hours of tedium and a meticulous attention to detail. But which details count, and which can be safely bypassed? Only a theory can tell us; only a theory can "suggest new, different and tractable ways to make observations."[11] Unfortunately, theories can have another, less benign effect; they can also "act as straitjackets to channel observations toward their support and forestall potentially refuting data."[12] Gould feels that this is the case with three firm tenets of Darwin's theory, the assumption that evolution is always gradual, that contingency has no serious role to play, that any gaps in the fossil record will eventually be filled in.

It was this brick wall of orthodoxy that greeted a team led by physics Nobel Prize–winner Luis Alvarez. They claimed that the 20-million-year reign of the dinosaurs ended when an asteroid a few miles in diameter slammed into the earth, creating climatic chaos and geological mayhem, a surprise important enough to change the course of evolutionary history, and bizarre enough to be greeted with incredulity. Alvarez's is a story of how a scientific theory goes from bizarre to believable.

The tale begins in 1977; it ends on March 5, 2010, when *Science* publishes a report confirming his conjecture: "The correlation between impact-driven ejecta and paleontologically defined extinctions at multiple locations around the globe leads us to conclude that the Chicxulub impact triggered the mass extinction that marks the boundary between the Mesozoic and Cenozoic eras ~65.5 million years ago. This conclusion is reinforced by the agreement of ecological extinction patterns with modeled environmental perturbations."[13] After a long journey, hypothesis had become fact. The proof lay in showing that species that eventually went extinct flourished *right up to* the Cretaceous-Tertiary boundary. Testing this hypothesis involved a tedious stretch of fieldwork that had not up to that point seemed worth anyone's while.

Gould focuses on two researchers who undertook this apparently thankless task. Peter Ward started in the camp of disbelievers; meticulous fieldwork with fossil ammonites eventually forced him to change his mind. Peter Sheehan, a supporter from the start, decided to focus on the dinosaur extinction because he could recruit a small army of helpers from the Milwaukee Public Museum's "Dig-A-Dinosaur" program. These dedicated amateurs clocked 15,000 hours, the equivalent of nearly eight work-years, a massive effort that yielded the result Sheehan was seeking:

> Because there is no significant change between the lower, middle, and upper thirds of the formation, we reject the hypothesis that the dinosaurian part of the ecosystem was deteriorating during the latest Cretaceous. These finding are consistent with an abrupt extinction scenario.[14]

Gould and Niles Eldredge suggested a more radical version of this insight into nature's contingency, a serious modification of Darwinian orthodoxy. They averred that the gradualism that helped retard the acceptance of this sudden mass extinction was, in fact, a flawed inference from evolutionary theory. Their modification, punctuated equilibrium, insisted that the gaps in the fossil record, which Darwin had attributed to its imperfection, were real and permanent. They were evidence of stasis, of evolution standing still. Over millions of years, most species were relatively stable in form and genetic makeup. At some point, however, a small population permanently separated from the rest and underwent rapid evolutionary change—became a new species in thousands rather than millions of years. This was not evolution at the level of the individual; it was evolution at the level of the species.

Punctuated equilibrium, the subject of several of Gould's essays, is not a theory like special relativity, quantum electrodynamics, or molecular genetics or sociobiology, an idea designed to transform or establish a discipline; rather, it is a theory that revisits a central idea of the modern synthesis of evolutionary biology—pervasive gradualism—and finds it wanting. We may doubt its truth but we cannot doubt its importance, as measured by the size of the fuss it is still making. A formidable authority in favor of punctuated equilibrium is worth quoting, the legendary biologist Ernst Mayr:

> What had not been realized before is how truly Darwinian speciational evolution is. It was generally recognized that regular variational evolution in the Darwinian sense takes place at the level of the individual and population, but that a similar variational evolution occurs at the level of species was generally ignored. Transformational evolution of species (phyletic gradualism) is not nearly as important in evolution as the production of a rich diversity of species and the establishment of evolutionary advance by selection among these species. In other words, speciational evolution is Darwinian evolution at a higher hierarchical level. The importance of this insight can hardly be exaggerated.[15]

Given the importance of punctuated equilibrium to Gould—he clearly regarded it as central to his legacy—we would expect that the theory would be a prominent topic in his essays. We might also expect that in its defense Gould would employ the superior literary skills even his enemies acknowledge. He might do what Robert Wright accuses him of: "All told, Gould probably commands the largest and most enthusiastic readership of any evolutionist in this century. But within one small audience, the cheers are muted. A number of evolutionary biologists complain—to each other, or to journalists off the record—that Gould has warped the public perception of their field."[16] In a topic in which he had a strong personal investment, he might break his promise to his readers: he might distort and exaggerate, he might create a false dawn.

Not so. The four essays on punctuated equilibrium that Gould thought good enough to republish in book form exhibit neither special pleading nor flawed reasoning, only surprising inferences supported by facts, a chain that permits readers to share his experience of discovery. In "Lucy on the Earth in Stasis," Gould reports on a new fossil find, an *Australopithecus afarensis* skull far younger than the famous Lucy, a fossil dubbed "Son of Lucy":

> The new skull, at 3.0 million years old, represents the youngest known material of *A. afarensis*. Since the bones are indistinguishable from skull pieces found earlier among the older specimens, Lucy's "son" demonstrates nearly half a million years of stasis in the first species of our distinctive evolutionary bush.[17]

To make his point about the pervasiveness of evolutionary orthodoxy in the interpretation of this fossil find, Gould cites the *Christian Science Monitor* writer R. C. Cowan: "What's remarkable about this 3-million-year old fossil is not that it is so old but that it's so young. It is 200,000 years younger than the famous Lucy . . . and a million years younger than the oldest specimen. Yet it looks like those ancestors."[18] To Gould, Cowan's surprise comes only from the irrational persistence of a bias in favor of gradual change, a bias Cowan shares with most of his readers.

Those subject to this pervasive bias are persistently puzzled by the fact that human beings do not seem to be gradually evolving. Ours seems to be a steady state. Is the cause cultural constraints that have undermined the ruthless force of natural selection? Is this evolutionary stagnation? Anyone who believed with Gould that stasis is the norm, change an infrequent occurrence, would refrain from asking such questions. They would cease to be surprised and would view with admiration a theory that makes better sense of the evidence.

Figure 9.4 makes Gould's case. It is a lineage that "proposes as many as seven branching events within a restricted interval following Lucy's demise—a period shorter than the interval of Lucy's own stasis."[19] What looks anomalous to Cowan is just business as usual, stasis followed by a burst of activity. In this essay, we have not merely learned science; we have experienced what it is like to be a scientist.

"In the Mind of the Beholder" has a target far broader than the defense of punctuated equilibrium. Gould wants his readers to follow him as he links three events whose conjunction anyone might find surprising: the Cambrian explosion in speciation; the single point of origin for humans,

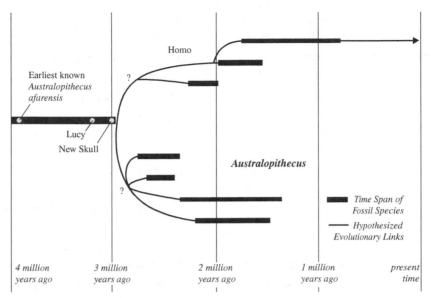

FIGURE 9.4 From Lucy to us.

SOURCE: Stephen Jay Gould, *Dinosaur in a Haystack* (New York: Harmony, 1995), 143.

probably in southern Africa; and the arrival of pollinating insects well before the flowers that are, supposedly, so essential for pollination. While punctuation can explain the first two events, the third seems an outlier. Yet for Gould all three events are examples of the same broader phenomenon. Evolutionary biology is a historical science. Like history, it is characterized by contingency, the same contingency that characterized the Panda's thumb and the QWERTY keyboard: "The newly shortened Cambrian explosion is decidedly unstately, the independence of insects and flowers seems chaotically uncoordinated, and the emergence of *Homo sapiens*, if viewed as a historical event in a single place, becomes quirky and chancy."[20]

In the final two essays of my set, evolutionary theory is a springboard to surprising conclusions. "Ten Thousand Acts of Kindness" begins with anecdotes about visitors who pilfer fossils under the mistaken impression that they valuable. In fact, they are worthless, the natural consequence of long periods of evolutionary stasis, punctuated by the rare events of speciation that make evolutionary history: "If we tried to infer the nature of species from the process that constructs the history of life, we would get everything precisely backward!—for events of great rarity (but with extensive consequences) make history."[21]

Still, we tend to act as if history, which consists largely of "warfare, greed, lust for power, hatred, and xenophobia,"[22] accurately reflects our everyday experience. In fact, "thousands of tiny and insignificant acts of kindness and consideration" characterize our daily experience: "We step aside to let someone pass, smile at a child, chat aimlessly with an acquaintance or even with a stranger."[23] This is not to deny the impressive power of violence in shaping our world: "One supposed insult, one crazed act of assassination can undo decades of patient diplomacy, cultural exchanges, peace corps, pen pals."[24] To Gould, "the solution to our woes lies not in overcoming our 'nature' but . . . allowing our ordinary propensities to direct our lives."[25] This view of a dilemma central to our human nature is a consequence of Gould's world view, a perspective that allows us to reflect deeply on the tension between permanence and change, between the worthy and the newsworthy.

In "Cordelia's Dilemma," the final essay of the set, Gould reverses the direction of our surprise: he moves from human nature to the way science is communicated, from the way science is communicated to the way we think of evolutionary theory. He begins with *King Lear.* Rather than flatter her father, as her sisters do, Cordelia prefers a silence that has its source in genuine affection. Lear misinterprets Cordelia's behavior: he takes the absence of evidence as evidence of absence. What is the link between Shakespeare and science? It is the tendency to prefer positive over negative results, a form of behavior that distorts the published record of science. Anne Fausto-Sterling shows how such bias works. In studies of differences between cognitive and emotional styles in men and women, she

> does not deny that genuine differences often exist, and in the direction conventionally reported. But she then, so to speak, surveys her colleagues' file drawers for studies not published, or for negative results published and then ignored, and often finds that a great majority report either a smaller and insignificant disparity between sexes, or find no difference at all. When all studies, rather than only those published, are collated, the much-vaunted differences often devolve into triviality.[26]

While there is no easy fix for publication bias, a solution does exist within the community of science: Fausto-Sterling's kind of scrutiny.

There is another kind of bias, however, that resists resolution from within, one that existing theory commitments create. In Gould's view, it is this bias that accounts for the resistance he and Eldredge experienced when they advocated for the theory of punctuated equilibrium. Central to their advocacy was their insistence that the absence of evolutionary change over long periods of time required careful study, that this silence, this nothing, was significant in any consideration of the origin of species, that "stasis is data,"[27] a surprising fact that licenses a change the way we view our evolutionary past.

Experiencing the History of Science

As we flip through the science section of the *New York Times* Tuesday after Tuesday, it seems as if a great deal of current science comes and goes. Still, we might find it hard to believe that the periodic table and the germ theory of disease, having had their day, will pass from the scene. The problem is that we won't be around to see if we are right. The only way to gain that kind of perspective is to turn the clock back. Philosopher R. G. Collingwood tells us how:

> As natural science finds its proper method when the scientist, in Bacon's metaphor, puts Nature to the question, tortures her by experiment in order to wring from her answers to his own questions, so history finds its proper method when the historian puts his authorities in the witness-box, and by cross-examining extorts from them information which in their original statements they have withheld, either because they did not wish to give it or because they did not possess it.[28]

In his essays on the history of science, Gould's goal is parallel to that of his scientific work: to reconstruct the past article by article, book by book, just as a paleontologist reconstructs a dinosaur skeleton from its scattered remains. Buridan's impetus theory, Ptolemy's dismissal of terrestrial rotation, Darwin's pangenesis—what accounts for these examples of discarded science? Clearly, we cannot dismiss Buridan, Ptolemy, and Darwin out of hand. They were, without question, good, even great scientists. To resolve this paradox—the paradox that good scientists get things wrong—is the historian's task. Only the study of history reveals just how and why these surprising reversals of fortune occur, why theory turns into trash and trash into theory.

In "The Upwardly Mobile Fossils of Leonardo's Living Earth" Gould discusses da Vinci's mistaken theory of the earth. When Leonardo arrived on the scene, two rival theories were in place. The first was Noah's flood, a singular cataclysm that could not account for sequential fossil placement; the second consisted of Neoplatonic accounts in which fossil shells grew within the rocks that surround them; they are geological, not organic

formations. This could not be right either. If shells added growth rings—
and we know they did—the rocks within which they resided would have
fractured, would they not? To Leonardo, such mistaken explanations must
be abandoned in favor of a theory in which the earth, the macrocosm, imitates man, the microcosm:

> Man has been called by the ancients a lesser world, and indeed the term is
> rightly applied, seeing that if man is compounded of earth, water, air, and
> fire, this body of the earth is the same; and as man has within himself bones
> as a stay and framework for the flesh, so the world has rocks which are the
> supporters of the earth; as man has within him a pool of blood wherein the
> lungs as he breathes expand and contract, so the body of the earth has its
> ocean, which also rises and falls every six hours with the breathing of the
> world [the tides]; as from the said pool of blood proceed the veins which
> spread their branches through the human body, in just the same manner the
> ocean fills the body of the earth with an infinite number of veins of water.[29]

For this theory to be correct, Leonardo needed to show that the earth's
water circulated like the body's blood. Understandably, he failed to demonstrate any such thing: water does not climb uphill. But he did prove
to his satisfaction that the earth experienced an uplift: marine fossils far
above sea level provided the evidence. Gould shows us just why and just
how Leonardo got his science wrong and right. He does so by the sympathetic reconstruction of the circumstances in which the theory arose. We
are with him as he plies the historian's trade in search of the surprise it
generates, the revelation that it is difficult indeed to get science just right,
to reach the authentic scientific sublime.

Gould also plies the historian's trade in "Linnaeus's Luck," an essay
about a theory of taxonomy that turned out to be right in one realm, wrong
in another. Linnaeus's binomial system for classifying living things was
invented under the sway of creationist theories. Yet it remains robust to
this day. We join Gould as he wends his way to the surprising solution to
this puzzle: "The very structure of a binomial name encodes the essential
property that makes Linnaeus's system consistent with life's evolutionary
topography," namely, evolutionary branching.[30] Because of this convergence, Linnaeus's system and evolutionary theory are in sync; in both, we
start with the kingdom, say, Animalia; we move on to a branch, the phylum
Chordata; then to a branch of Chordata, the class, Mammalia; then to a
branch of Mammalia, the order Carnivora; then to a branch of Carnivora,
the genus *Canis*; then to the species *Canis familiaris*, the domestic dog,

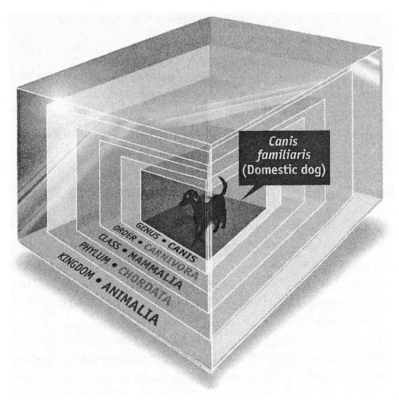

FIGURE 9.5 Classifying your household pet.

SOURCE: Stephen Jay Gould, *I Have Landed: The End of a Beginning in Natural History* (New York: Harmony, 2002), 288.

then to Lassie. In figure 9.5 Gould employs a visual that makes the system concrete and memorable.

Gould is not entirely happy with a solution that depends solely on Linnaeus's perception that "under the logic of hierarchy and branching, organisms could be arranged into a consistent order that might win general consent without constant bickering among practitioners."[31] He feels that Linnaeus was also tapping into something more fundamental, the way our minds work:

> Linnaeus also succeeded because he made a very clever and probably con-
> scious choice from the mental side of taxonomic requirements as well. In
> deciding to erect a hierarchical order based on continuous branching with
> no subsequent joining of branches, Linnaeus constructed his system accord-
> ing to the most familiar organizing device of Western logic since Aristotle
> (and, arguably, an expression of our innate and universal mental preferences

as well): successive (and exceptionless) dichotomous branching as a system for making ever finer distinctions.[32]

This explanation is of interest because the way the mind works also led Linnaeus astray. His application of the binomial system to minerals and diseases failed because a mental predisposition rode roughshod over the facts on the ground. There was another potentially more devastating surprise in store for Linnaeus's system: lateral gene transfer, a process that could undermine it. By this means, bacteria can violate the dichotomous principle of exceptionless branching. Gould quotes a leading researcher:

> If "lateral gene transfer" can't be dismissed as trivial in extent or limited to species categories of genes, then no universal hierarchical classification can be taken as natural. Molecular phylogeneticists will have failed to find the "true tree," not because their methods are inadequate or because they have chosen the wrong genes, but because the history of life cannot properly be represented as a tree.[33]

It is just possible that current scientists are the victims of an inbred mental predisposition that misled a great scientist centuries ago.

In "Brotherhood by Inversion (or, As the Worm Turns)," Gould shows us that scientists must be cautious in ridiculing the apparently ridiculous: a theory that seem wildly wrong at first may, surprisingly, turn out to be right at last. For centuries, scientists wondered whether an evolutionary relationship existed between vertebrates and invertebrates. While many structural aspects seemed alike, central nervous systems were starkly opposed: in vertebrates, the system lies above the gut; in invertebrates, below. If only we could show how such a shift in position might occur, we would have removed a stumbling block in the way a single line of evolutionary advance from microorganisms to Madonna.

In "Brotherhood," Gould introduces us to Walter H. Gaskell, an otherwise solid scientist who championed what seemed to be a truly dotty idea. According to Gaskell, the invertebrate gut was transformed into the vertebrate central nervous system and a new gut simply evolved. Figure 9.6, an illustration from Gaskill, shows how unlikely such a development would be, given the relative positions of the central nervous systems.

While theories of evolutionary continuity have suffered from a deficiency of evidence, they derive from a long and honorable tradition that attempts "to reduce organic diversity to one or a very few archetypical building blocks that could then generate all actual anatomies as products

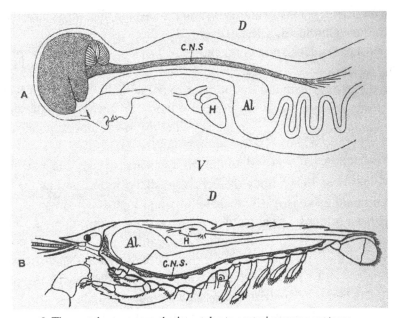

FIGURE 9.6 The vertebrate versus the invertebrate central nervous systems.

SOURCE: Stephen Jay Gould, *Leonardo's Mountain of Clams and the Diet of Worms: Essays on Natural History* (New York: Three Rivers, 1998). From Walter Holbrook Gaskell, *The Origin of the Vertebrates* (New York: Longmans, Green, 1908), 10.

of rational laws of transformation."[34] Prominent among their advocates was Étienne Geoffroy Saint-Hilaire, an early 19th-century scientist interested in "establishing a 'unity of type' that could generate both arthropods [like insects and lobsters] and vertebrates."[35] So the tradition within which Gaskell was working was perfectly respectable; moreover, Gould shows us that Gaskell's theory was no more off the wall than the theory already in place, the idea that invertebrates simply "flipped over" to form their vertebrate brothers.

Respectable does not equal acceptable. Although Geoffroy's theory was not laughed out of court, it did not survive. Mainstream evolutionary biology endorsed the idea that "arthropods and vertebrates do not share the same anatomical plan; instead, they represent two separate evolutionary developments of similar complexity from a much simpler common ancestor that grew neither a discrete gut nor a central nervous chord."[36] This very sensible orthodoxy, as Ernst Mayr astutely observed, precluded the genetic conditions that would be the basis of a unity of type: "In the early days of Mendelism there was much search for homologous genes that would account for such similarities [in design]. Much that has been learned about gene physiology makes it evident that the search for homologous genes is quite futile except in very close relatives."[37] Despite this stern and sensible admonition, Charles B. Kimmel engaged in the search that Mayr regarded as quixotic. Prudence and common sense lost out: Geoffroy was on the right track after all. There are homologous genes. Indeed, Kimmel tells us that "we have come to find it more remarkable to learn that a homolog of our favorite regulatory gene in a mouse is not, in fact, present in *Drosophila* than if it is, given the large degree of evolutionary conservation in developmentally acting genes."[38]

In their understanding of nature, past scientists were constrained by their own natures, by strong theoretical predispositions, and by the general culture they shared with their fellows and their fellow citizens. How can present scientists be any different? How can they avoid being victims of hidden biases that undermine their deepest insights, their most cogent inferences, their theories that lead, they hope, to an authentic sense of wonder evoked by the true scientific sublime? They cannot. The scientific future will be full of surprises, pleasant and unpleasant, discoveries touching the sublime, "discoveries" off by a mile.

Conclusion

While in his professional work, Gould's voice is consistently that of science, his popular essays exhibit not one voice but five: the scientist, the historian, the literary critic, the music lover, and the raconteur. Together, these constitute the character "Stephen Jay Gould," the creation of Stephen Jay Gould, literary artist, the heir of Michel de Montaigne. In the essays just surveyed, a genial tour guide engages us by surprising us, inviting us to join him in a voyage of discovery as he conceals and then reveals insights into the wonders of nature, the patterns that science discovers and, having discovered, explains as the consequence of nature's laws.

CHAPTER 10 | Steven Pinker

The Polymath Sublime

In the beginning was the Word, and the Word was with God, and the
Word was God.

—*John* 1.1

ALREADY FAMOUS AT 40, Josef Haydn was searching for new means of
expression. The result was his six Opus 20 string quartets, a dazzling set
whose new directions put their stamp on every composer who has since
attempted the form. For those accustomed to previous quartets, including Haydn's own, every minor turn was a major surprise, each new direction conveying a sense of the composer's joy as he reveled in his mastery
of his medium. At 40, already a well-respected cognitive scientist at the
Massachusetts Institute of Technology, Steven Pinker came suddenly to
the world's attention with his first book of popular science, the bestselling
Language Instinct, an embodiment of the linguistic sublime. Emboldened
by instant fame, he followed this achievement by following Francis Bacon,
making all knowledge his province, telling us how the mind works, why it
isn't a blank slate, and why violence has declined.

Not many professors are interviewed by Stephen Colbert; not many can
be described as a brilliant lecturer who looks like a rock star: "His curly
shoulder-length mane and Cuban heels give him the air of a prog rocker
on his third comeback tour. He has a superbly defined jaw, glittering blue
eyes and a kilowatt smile which he beams at his class as he switches on
the microphone."[1] Not many professors find themselves on a poster that
updates Raphael's famous painting *The School of Athens*, a gathering of
ancient worthies. Figure 10.1 is a depiction easily identified by the carica-
ture's flowing locks.

FIGURE 10.1 Guess who? *Lingua Franca* poster.

Raised by middle-class Jewish parents in Montreal, Pinker first distinguished himself as the graduate student of the prominent Harvard psychologist Steven Kosslyn, who said of him: "He was officially my student, but almost from the start we were colleagues."[2] After studying with Kosslyn, Pinker went on to carve out a successful academic career as an experimental psychologist, first at MIT, then at Harvard, specializing in language acquisition in children. But he was not satisfied as a mere academic star, much sought-after, much honored, destined to shine brightly but not to dazzle. Accordingly, he chose another firmament, science popularization, anxious to become its North Star. In the words of his friend and fellow professor Jay Keyser: "He was a careful worker, obviously a player. Then came *The Language Instinct* and he suddenly found a voice that may have been buried in his work but which I hadn't seen before. And it was superb. It was the perfect tone about a really complicated field to an intelligent lay audience that had not been addressed."[3]

A review of *The Language Instinct* in the leading scientific periodical, *Nature*, was typical: "Here at last is a marvelously readable book about language, written by a real expert. Steven Pinker tackles with wit and erudition the kinds of question everyone asks about language."[4] Pinker himself explains the origin of this miraculous entry onto the world's stage:

I've always enjoyed the challenge of conveying difficult concepts without dumbing them down. I've long been a teacher (I put myself through college as a math tutor and Jewish Sunday-school teacher), and have long paid attention to the mechanics of pedagogy—especially the use of language, visuals, and analogies to get ideas across. Not only has my research touched on all of these mechanisms (I once was involved in a project on the perception of graphs, for instance), but I read style manuals for fun, and like to analyze sentences I admire in other people's prose to figure out why they work. I have no way of knowing whether I inherited any talent at communication—my parents and siblings are highly articulate, but of course I grew up with them, so we don't have an unconfounded nature-nurture comparison there. But whatever talents I did happen to be born with, I certainly cultivated, and continue to cultivate.[5]

His subsequent productivity is indeed impressive: "When I'm writing a book, I can't rest until I finish. I will typically write until 3 in the morning, seven days a week, until I'm done. I'll try to set off blocks of time, like a sabbatical or a summer, in which I can get up in the morning and write till I go to bed. I worked on *How the Mind Works* for virtually an entire year. I took off a month to get married."[6] He relaxes by bicycling and dabbling in photography. "When?" one wonders. *How the Mind Works* weighs in at 659 pages.

I will follow this polymath on a journey that will give us the opportunity to differentiate among the ways in which he employs scientific sublimity. I will distinguish his evocation of wonder at patterns of nature hidden in plain sight from his evocation of wonder at their explanation. I will explore instances in which general readers are asked to side with Pinker in his quarrels with other professionals, a dubious maneuver, it seems to me, at odds with the practice of disciplines. Finally, I will analyze the degree to which science can extend its reach to our mental capacities, our culture, and our history. In these cases, our exploration will allow us to differentiate arguments that infuse science with a sense of awe from those that infuse mere opinion with scientific authority.

Pinker as Linguist

In the latter half of the tenth century, the Anglo-Saxon monk Aelfic wrote a homily on daily miracles, the power of the God who sustains the universe moment by moment, a marvel hidden in plain sight:

God hath wrought many miracles and daily works; but those miracles are much weakened in the sight of men, because they are very usual. A greater miracle it is that God Almighty every day feeds all the world, and directs the good, than that miracle was, that he filled five thousand men with five loaves: but men wondered at this, not because it was a greater miracle, but because it was unusual. Who now gives fruit to our fields, and multiplies the harvest from a few grains of corn, but he who multiplied the five loaves? The might was there in Christ's hands, and the five loaves were, as it were, seed, not sown in the earth, but multiplied by him who created the earth.[7]

Steven Pinker strives to be the Aelfric of linguistics, to explore with us the ordinary miracle of language acquisition and use. He labors successfully to make clear to the general public the insights of Noam Chomsky, the linguist who claims that the intricate patterns that characterize the world's languages, their differing surface structures, are the consequence of a common core, a universal grammar, the generator of all languages. The success of *The Language Instinct*, and of Pinker's two subsequent books on language, was due in part to the clarity of his exploration of this phenomenon. But it is not this clarity that makes his work compelling; it is his evocation of the scientific sublime, a revelation that shows us that something we all possess—something we all take for granted—is a daily miracle that rewards our scrutiny, an examination that reveals a human capacity that science lays bare and explains.

Pinker reveals linguistics as the science that shows us just how much of a miracle language is. All of us routinely turn its storehouse of words and rules into an endless number of sentences, many of which are new-minted for the occasion. Because we achieve these articulate heights by the age of three, we can't have learned language just from hearing our parents speak. We must have been primed in advance; each of us must have an all-purpose language generator packed inside our skulls. Of course, we all speak different languages: French children learn French, American children English. But this is only because a universal grammar, an innate capacity, made this differentiated learning possible. It is this grammar, modified to accommodate the rules of particular languages, that generates the sentences we speak. If American, we learn that almost all adjectives precede their nouns. Americans say, "the blue ink." If French, we learn that adjectives generally follow their nouns. The French say "l'encre bleu." Modern linguistics tells us that these differences obscure commonalities, persistent patterns that are themselves a source of wonder; it also reveals the generator of these patterns, their explanation, a source of wonder even more impressive than the first.

Modern linguistics deals not only with language but with speech. If we had language but not speech, we could still communicate effectively: the deaf do it every day. But most of us rely on another miracle: our ability to convey our thoughts through the sounds we make. This capacity to form sounds into words and sentences and to perceive those sounds as sentences evolved in coordination with the universal grammar, a simultaneous development so impressive it gives some philosophers pause when evolution is given as their explanation. What on earth could have been the selection pressures involved?

Even Pinker, a man who will insist that he knows how the mind works, doubts our ability to understand human speech recognition: "The human brain . . . is a high-tech speech recognizer, but no one knows how it succeeds";[8] in fact, "no system today can duplicate a person's ability to recognize both many words and many speakers."[9] Our speech generator is equally impressive:

> First, one of the six speech organs is chosen as the major articulator: the larynx, the soft palate, tongue body, tongue tip, tongue root, or lips. Second, the manner of moving that articulator is selected: fricative, stop, or vowel. Third, configurations of the other speech organs can be specified: for the soft palate, nasal or not; for the larynx, voiced or not; for the tongue root, tense or lax; for the lips rounded or unrounded. Each manner or configuration is a symbol for a set of commands to the speech muscles, and such symbols are called features. To articulate a phoneme, the commands must be executed with precise timing, the most complicated gymnastics we are called upon to perform.[10]

That a three-year-old can have mastered this mental and physical feat is truly astonishing.

Each language has its phonetic fingerprint, the growl of German, the lilt of French. This is how the great comedian Sid Caesar seemed to speak German or French, while speaking nonsense. Pinker quotes a woman who produced under hypnosis the following pseudo-Slavic nonsense: "Ovishta reshra rovishta. Vishna beretishti? Ushna barishta dashto."[11] Once the shape of a language is firmly embedded, however, the pattern is resistant to change. Every Chinese child speaks flawless Chinese. But few American adults can learn to speak flawless Chinese: almost all speak it with an American accent. Although the language is clearly Chinese, the pattern of sounds the speaker generates is tainted indelibly by his native tongue. "Here is Jack and the Beanstalk" with an Italian accent: "Uans appona taim uase disse boi. Neimmse Giacche. Naise boi. Live uite ise mamma. Mainde de cao."[12] While most adults cannot go from language to language and sound like natives, many actors can seem to. In the ancient radio show *Life with Luigi*, J. Carrol Naish and Alan Reed adopted an Italian accent, transforming "Jack and the Beanstalk" into "Giacche enne Binnestauche."

Pinker tells us that just as there are rules for grammar, there are rules that govern phonology, determining how features are arrayed into phonemes and words. In the written language, many verbs end in "ed" in the past tense: slapped, jogged, bedded. While these endings are spelled the same way, they do not sound the same way. In "slapped" the "ed" is pronounced "t"; in "jogged" it is pronounced "d"; in "bedded" it is pronounced "ed." In each case, the reason is phonological, and a rule applies. After an unvoiced vowel, we get "t"; after a voiced vowel, we get "d"; after a "d" (or a "t"), we get "ed." In each case, the constraints on our speech apparatus determine our pronunciation.

But if this is so, why do we persist in tolerating the gap between pronunciation and spelling? Why not spell words just the way they sound: slapt, jogd, bedded? Isn't English spelling irrational? Certainly, it does not make sense to have a "b" in *debt* or a "k" and a "gh" in *knight*. They are there only because history put them there, the "b" because of a mistaken analogy with Latin, the "k" and "gh" because these letters were once pronounced. The arrival of dictionaries regularized spelling and fixed these oddities in place, seemingly forever. As Pinker points out, however, phonetic spelling would only make things worse. We would spell "courage" differently from "courageous," "muscle" differently from "muscular." The fact is that "writing systems do not aim to represent the actual sounds of talking, which we do not hear, but the abstract units of language underlying them, which we do hear."[13]

The Language Experts

In "The Language Mavens," a chapter of *The Language Instinct*, Pinker makes the point, well supported by linguistic science, that "common people, no matter how poorly educated, obey sophisticated grammatical laws, and can express themselves with a vigor and grace that captivates those who listen seriously—linguists, journalists, oral historians, novelists with an ear for dialogue."[14] From this perspective, he admonishes the self-appointed language expert John Simon for dismissing Black Vernacular English as ungrammatical. From the same perspective, he corrects another maven, William Safire, who insists that George W. Bush was ungrammatical when he said "*Who* do you trust?" rather than "whom." Acknowledging that the latter seems stilted, Safire suggests an alternative: Which candidate do you trust? Pinker rightly points out that this solution, applied generally, leads to such abominations as "Abe, which child or children did you play with today at school?" Pinker does acknowledge feeling queasy when *disinterested* is used to mean *uninterested*. After all, *disinterested* is not really a synonym for *impartial* or *unbiased*. It means having no stake at all in the matter. In objecting to this misuse, Pinker seems to be behaving as a language maven himself. Although this loss of meaning impoverishes English irretrievably, Pinker takes comfort in the fact that language inevitably changes, its riches constantly being replenished. Yiddish has enriched the English language, for example. Witness *maven*, Yiddish for *expert*.

It is beyond question that linguistic science supports the legitimacy of dialects like Black English. But there is a difficulty with Pinker's attack on the language mavens. He complains that they do not understand linguistic science, a deficiency that undermines any claims they may make about linguistic propriety. The language mavens, however, are interested not in the science of language but in its optimal employment. It is no criticism of Black English to say that while it is perfectly suited to intimate encounters in the African American community (just as Yiddish is perfectly suited to such encounters with the Hassidic community), it is not suited to the wide range of uses of the standard dialect in science, scholarship, and the literary arts. The language mavens are the Emily Posts of language. Just as Emily Post was concerned with how people should behave if social friction is to be minimalized, the language mavens are concerned with how writers should behave if language functions as an art in which misunderstanding is avoided, ungainliness eliminated, and grace abounds. To accomplish these aims, careful writers have forged versions of the standard dialect that do more than obey the rules of a universal grammar adjusted to the constraints of a natural language. These writers have created enhanced dialects with a power and flexibility appropriate to the tasks at hand. For them, language is an art, a deliberate search for *le mot juste*, for precision and eloquence joined at the hip. It is this quality that John Simon and William Safire champion; it is also this quality that they exemplify. And it is this quality, it must be said, that Steven Pinker also exemplifies.

Pinker Agonistes

Noam Chomsky is the most influential advocate of the theory that language is generated by rules built into our brains. His 1957 *Syntactic Structures* created a revolution in the study of language. Two years later, in a devastating review of B. F. Skinner's behaviorist approach to language in his *Verbal Behavior*, Chomsky reminded us of his alternative, one that would prove fruitful indeed. Chomsky sees the grammar of a language as a mechanism that generates sentences in the way that a deductive theory generates theorems. Linguistics can be viewed as a study of the formal properties of such grammars. With a precise enough formulation, this general theory can provide a uniform method for determining a structural description that can give us an insight into how any sentence is used and understood. In short, it should be possible to derive from a properly formulated grammar a statement of the integrative processes and generalized patterns imposed on the specific acts that constitute an utterance.[15] According to Chomsky, the different rules of English or Japanese are shaped by the constraints of a universal grammar, a common source of all languages.

In 1986, David Rumelhart and James McClelland proposed an alternative to a rule-based system. There were rules of a sort, such as the rule that most past tenses in English end in "ed." But these rules were generated by the experience of children with the spoken language, an experience represented in neural networks by an increase in the probability of the "ed" ending, the ending of all regular verbs. Of course, irregular verbs also must be learned. Some come in patterns: for example, slept, kept, wept, and crept; a few others, frequently occurring oddballs, such as, *was, go, went*, must be learned by rote.

In this research program, connectionism, language learning is just learning. Rumelhart and McClelland summarize their position:

We have, we believe, provided a distinct alternative to the view that children learn the rules of English past-tense formation in any explicit sense. We have shown that a reasonable account of the acquisition of past tense can be provided without recourse to the notion of a rule as anything more than a description of the language. We have shown that for this case, there is no induction problem. The child need not figure out what the rules are, nor even that there are rules. The child need not decide whether a verb is regular or irregular. There is no question as to whether the inflected form should be stored directly in the lexicon or derived from more general principles. There isn't even a question (as far as generating the past-tense form is concerned) as to whether a verb form is one encountered many times or one that is being generated for the first time. A uniform procedure is applied for producing the past-tense form in every case. The base form is supplied as input to the past-tense network and the resulting pattern of activation is interpreted as a phonological representation of the past form of that verb. This is the procedure whether the verb is regular or irregular, familiar or novel.[16]

Obviously, these views are a challenge to any rule-based system. Pinker's counter strategy points out the many problems Rumelhart and McClelland have in accounting for every instance of the past tense. For example, because their model is based exclusively on the sound of words, it cannot account for those that sound alike but have different plurals, words such as wring and ring. Given this and related difficulties, Pinker feels justified in dismissing "the knee-jerk associations that drive the Rumelhart-McClelland model."[17] This is the position of Pinker and Alan Prince, whose article dismantles this model, piece by piece. Nevertheless, it turns out, paradoxically, that Pinker and Prince "are probably the model's biggest fans."[18] In *Words and Rules*, Pinker proposes a compromise that, while it applies the Rumelhart-McClelland model to all irregular verbs, applies his own rule-based model to all regular ones. For Pinker, there must be rules for regular verbs.

McClelland is having none of it. There may very well be rules, but they would not have been there in the first place; they would have developed as the outcomes of probabilistic processes:

> We do not claim that it would be impossible to construct a rule-based model of inflection formation that has all of the properties supported by the evidence. However, such an account would not be an instantiation of Pinker's symbolic rule account. In fact, rule-based models with some of the right characteristics are currently being pursued. . . . If such models use graded rule activations and probabilistic outcomes, allow rules to strengthen gradually with experience, incorporate semantic and phonological constraints, and use rules within a mechanism that also incorporates word-specific information, they could become empirically indistinguishable from a connectionist account. Such models might be viewed as characterizing an underlying connectionist processing system at a higher level of analysis, with rules providing descriptive summaries of the regularities captured in the network's connections.[19]

Connectionist theory is not something that can be cavalierly dismissed by Pinker's suggestion of political compromise. Like Chomsky's *Syntactic Structures*, Rumelhart and McClelland's initial article has initiated a tsunami of empirical research, created by an army of men and women staking their careers on the fruitfulness of connectionist theory. From their point of view, the rule-based program of Chomsky and Pinker may be worse than wrong; it may inhibit our understanding of the nature of language; it may mislead. Linguist Franklin Chang concludes that

> it is possible that the requirement that syntactic knowledge in different lan-
> guages be instantiated in a common syntactic theory actually hinders the
> development of explicit models. Ironically, by making fewer assumptions
> about the universal nature of syntactic knowledge and allowing learning to
> determine the constraints for each language, we might actually get closer to
> a universal account of human language behavior.[20]

Pinker's treatment of his differences with those of well-respected philosopher Jerry Fodor parallels his treatment of Rumelhart and McClelland. Pinker ridicules Fodor's view of the origin of concepts, comparing him to a fictitious Manny Schwartz, the inventor of the Theory of Continental Drip, the contention that the shape of the southern continents may be attributed to their initial molten state.[21] In contrast with this attempt at character assassination, listen to what philosopher Stephen Schiffer says about Fodor, whose views on language he cannot abide. In a *TLS* review of *Concepts*, Schiffer speaks of "the richness of considerations adduced by Fodor in this important, aggressively argued, accessible, witty, irreverent, wide-ranging, provocative, and bound-to-be-influential book written in his famous, no-niceties, shoot-from-the-hip style."[22]

The question facing Fodor and Pinker is the same: What is the origin of concepts, the meanings behind meanings? Clearly, we didn't enter the world with concepts like TROMBONE and ELECTRON inside our heads. On the other hand—and this is the problem—concepts can't be learned. Although we can learn *about* trombones and electrons, we can only do so by asking questions like: What does a trombone sound like? How big is an electron? But such questions assume that we *already* have the concept of TROMBONE and ELECTRON. One possibility—a possibility favored by Pinker—is that we are born with certain basic concepts already in our heads, concepts like CAUSE and EFFECT. These are innate. Concepts like TROMBONE and ELECTRON are acquired, the progeny of some combination of these basic concepts.

There is a problem with this approach, one that Fodor hammers home.[23] Pinker's solution requires that his repertory of innate concepts be sufficient to define exactly what we mean when we use trombone or electron in a sentence. But all definitions save the mathematical are as full of holes as a colander. Let's take the verb *paint*. Let's define *paint* as "cover a surface with paint." No, that won't do. If we kick a can of paint over on the floor, we haven't painted the floor; if Michelangelo is working on the Sistine ceiling, he is not painting it; he is painting a picture on it; if he dips his brush into a pot of paint, he is not painting the brush, but preparing to paint the ceiling. The problem is general. KILL can't be decomposed into CAUSE TO DIE. Although I can cause you to die by not calling an ambulance, I haven't killed you. This problem has quite a reach: philosopher Edmund Gettier showed conclusively that you can't define knowledge as justified true belief; in fact, you can't *define* knowledge at all.[24]

Pinker thinks he has an answer to these difficulties. Fodor has confused definitions with semantic representations. For example, the semantic representation of the verb *paint* would include the idea that *to paint* is the act of an agent whose goal is to cover a surface with paint. But this fix is no fix at all. We still haven't dealt with the difference between *I painted the ceiling* and *I painted a picture on the ceiling.* Moreover, what are we to say of *I painted the wound with mercurochrome?* What about the verb *to butter?* Does a person who butters spread a viscous, yellowish edible substance, not necessarily butter, on another substance? Suppose I use butter to soothe a burn? Am I buttering the burn?[25]

Nobody, it seems, favors Fodor's original position or his revised one, in which he abandons the view that TROMBONE and ELECTRON are innate. For this, he substitutes a mysterious process in which trombones and electrons imprint their concepts on our brains the way ducklings are imprinted from birth with the concept MOTHER. But Pinker's solution— alleged semantic representations—is also a non-starter. Indeed, Pinker seems aware of his failure. In *How the Mind Works*, he gives us an example that defies analysis in terms semantic representation. A bachelor is a man who has never married. So is Arthur, who is living with Alice, a bachelor? Is Bruce, who has arranged a marriage simply to avoid the draft? Is Charlie, who is seventeen and lives with his parents? Is David, who is also seventeen but on his own and living the life of a playboy? Are Eli and Edgar, homosexual lovers living together? Is Faisal with his three wives, a man in search of a fourth? Is Father Gregory?[26]

I have no opinion about the relative merits of these two robust research programs, each impelled by impressive insights into the nature of language. More to the point, I do not think I am entitled to an opinion. First, I lack the knowledge required, and second, even the required knowledge is now insufficient to achieve closure on the part of the participants. I can only hope that experts will eventually agree and that that agreement will stand the test of time, the only index we have that we have arrived at scientific truth. In the meantime, we have Chomsky and Pinker on the one hand and Rumelhart, McClelland, and Fodor on the other. In each case, Pinker's scientific sublime is generated not by means of a widely accepted theory but by an interesting hypothesis in which he firmly believes.

How Pinker's Mind Works

To Pinker, our entire mental capacity is essentially computational. This is not to say that we are walking around with a laptop between our ears; it is to say, rather, that the brain processes our sensations in a way analogous to the way a computer processes its 0s and 1s. A computational theory certainly seems to explain some remarkable characteristics of human vision. Take the problem of how people identify shapes rotated in space. According to one theory, people retain each separate orientation in memory; according to another, they are capable of rotating these shapes in their own mental space; according to a third, geon theory, orientation doesn't matter; identification is simply a matter of pattern matching. Pinker conducts an experiment to decide which theory is correct. He is a scientist, after all, not just a popularizer of science. Experimentation demonstrates that people use all three methods, depending on the circumstances. But Pinker is not finished. He also wants to give us the flavor of his discovery, the sudden insight into the way the visual system works, an experience he calls "my happiest moment as an experimenter."[27]

In image rotation, there is an interesting anomaly. While the degree of image tilt generally correlates with the time it takes to identify it, this is not true of their mirror images, a result that continues to baffle. Disregarding this persistent puzzle, Pinker and his graduate student start to write up their paper. In it, they speak of a different "strategy" employed for mirror images, a sure indication that they are "clueless" about the real solution to their problem. It is then that "an idea hit[s]":[28] it turns out that a two-dimensional shape, rotated around its axis like a chicken on a spit, can always be aligned with its mirror image: the degree of tilt makes no difference. The computational theory of mind chalks up another victory.

Viewing the mind as a computer is fruitful but problematic. Pinker is not satisfied merely to explain feats of human vision. He wants also to trace them back to their origin. Uniformly, he attributes the origin of the incredible feats of which the human eye is capable to evolutionary forces, selection pressures operating on human variation, a process that begins with our simian ancestors and ends with us. An example is cyclopean vision—our ability to see with both eyes shapes that cannot be seen with only one:

> Primates evolved in trees and had to negotiate a network of branches masked by a veil of foliage. The price of failure was a long drop to the forest floor below. Building a stereo computer into these two-eyed creatures must have been irresistible to natural selection, but it could have worked only if the disparities were calculated over thousands of bits of visual texture.[29]

It is no dismissal of evolutionary theory to say that in this passage it is transformed into an ideology by means of which Pinker, in despair of evidence, concocts a blatant just-so story. There is no evidence for the truth of his conjecture. How could there be?

Pinker's view that the mind is the product of an evolution that is wholly materialistic is also problematic. Although he seems generally blind to the unfortunate implications of this view, nevertheless he puts his finger directly on their cause, the problem of consciousness, without which morality is impossible to imagine:

> No account of the causal effects of the cingulate sulcus [an area of the brain] can explain how choices *are not caused at all*, hence something we can be held responsible for. Theories of evolution of the moral sense can explain why we condemn evil acts against ourselves and our kith and kin, but cannot explain the conviction, as unshakable as our grasp of geometry, that some acts are inherently wrong, even if their net effects are neutral or beneficial to overall well-being.[30]

It is the philosopher Thomas Nagel who makes the appropriate inference: no theory of evolution can be correct that views it solely as a material process.[31] If we are conscious, and if consciousness is a consequence of our evolutionary history, then the precursor of consciousness in all its immateriality must have been present at the Big Bang: "Intentionality, thought, and action resist psycho-physical reduction and can exist only in the lives of beings that are capable of consciousness."[32]

There is a final problem, one with the computational theory of mind itself. Pinker, a dedicated advocate of this theory, sees the mind as an organ "packed with high-tech systems."[33] But surely Jerry Fodor is right to find this analogy between human and natural engineering less than compelling. How, he asks, can Pinker assert how hopeless we are when it comes to building a serviceable robot—one that can put away the dishes—and at the same time contend that we know that the mind works just like a robot's control system?[34]

Pinker does not have a theory of how the mind works; he has a model for the way some parts of the *brain* work. *How the Mind Works* champions a computation theory that is unable to explain why we can use a knife and fork and robots cannot, one that favors an evolutionary theory of mind that cannot explain how consciousness evolves from a primal soup of elementary particles.

The Blank Slate: Pinker on Gender

That men and women have varying talents is taken for granted; whether they have these simply as a consequence of being men and women is not. Only two women have won the Nobel Prize in physics; only one has won the Fields Medal, the highest honor in mathematics. Of the top 20 chess players of all time, not one is female.[35] While Clara Schumann was a good composer, her husband, Robert, was a great composer. While Lee Krasner was a competent artist, Jackson Pollock, her husband, was a leading abstractionist, the inventor of action painting. Have these trends been shaped largely by differences in the innate capacities of men and women or largely by social and political forces that have denied women the opportunity to achieve the highest levels?

To enter the public arena on these issues is to ignore prominently posted avalanche warnings. Just ask Larry Summers. At the National Bureau of Economic Research, the most prestigious economics research organization in America, an organization noted for its airing of unbiased views, this prominent economist spoke with candor. His topic was the effect of recent research in behavioral genetics on our perception of the gender gap. After his talk, he summed up the thrust of his argument:

> My point was simply that the field of behavioral genetics had a revolution in the last fifteen years, and the principal thrust of that revolution was the discovery that a large number of things that people thought were due to socialization weren't, and were in fact due to more intrinsic human nature, and that set of discoveries, it seemed to me, ought to influence the way one thought about other areas where there was a perception of the importance of socialization.[36]

The research Summers referred to suggested that the differential achievement of men and women in science, engineering, and mathematics, especially the low representation of women at the highest levels of these endeavors, might be attributed to differences in innate ability combined with a preference for careers that did not preclude the perfectly sensible option for lives that did not include 80-hour, seven-day work weeks. Summers stressed that these conclusions must be regarded as tentative and that it was open to question whether the social practices that created high-pressure careers benefited anyone, men or women. The feminist reaction was immediate, vitriolic, and wrong-headed:

> Summers's comments were, in fact, shockingly prejudiced to the detriment of women. This was NOT an instance in which feminists simply didn't want to hear an idea that might contradict our worldview. This was an instance in which the head of a major research university stated at an official event that he believes that women do not tend to advance to tenured positions at major research universities because (a) they don't WANT to put in the punishing 80 hour work weeks that are required and (b) genetic biological factors disfavor women's achievement at the highest levels in math and science. He stated that this was his opinion, even while also admitting that we don't actually KNOW whether this is the case. In other words, he WASN'T just broaching an idea for further investigation—he was endorsing a particular point of view regarding women's inferiority in a particular field while simultaneously admitting that he couldn't support his point of view.[37]

It is thought that the reaction to this speech contributed to Summers's resignation from his Harvard presidency and his failure to be appointed chair of the Federal Reserve.

Three years before Summers's notorious speech, in *The Blank Slate: The Modern Denial of Human Nature*,[38] Steven Pinker said essentially the same thing for the same reason. The facts on the ground seemed to support of claims of innate gender difference:

> Sex differences are not an arbitrary feature of Western culture, like the decision to drive on the left or on the right. In all human cultures, men and women are seen as having different natures. All cultures divide their labor by sex, with more responsibility for childrearing by women and more control of the public and political realms by men. (The division of labor emerged even in a culture where everyone had been committed to stamping it out, the Israeli kibbutz.) In all cultures men are more aggressive, more prone to stealing, more prone to lethal violence (including war), and more likely to woo, seduce, and trade favors for sex.[39]

These findings have nothing to do with the societal bias against women, about which Pinker has no doubt. He does not question that employers systematically underestimate the skills of women, that they worry unduly about women supervising men, or that they conclude without evidence that all-male workplaces are more efficient. But, he suggests, it is unlikely that gender bias is the cause of all differences between male and female achievement in the workplace. Speaking of the academy, an environment he knows best, Pinker asks whether it is plausible to believe that mathematicians place arbitrary barriers in the way of women, while evolutionary psychologists are fair-minded.[40]

Rape is the second hot-button issue with which Pinker deals. His first concern is to disabuse us of the notion that rape is about the oppression of women. In rejoinder, he poses a rhetorical question: "Most people agree that women have the right to say no at any point during sexual activity, and that if the man persists he is a rapist—but should we also believe that his motive has instantaneously changed from wanting sex to oppressing women?"[41] There is another problem. Without for a moment denying a young woman's right to say no even after getting drunk at a fraternity party and accompanying a young man to an upstairs bedroom, one may ask whether such conduct is prudent. Pinker quotes Camille Paglia on this point:

> "When you drive your car to New York City, do you leave your keys on the hood?" My point is that if your car is stolen after you do something like that, yes, the police should pursue the thief and he should be punished. But at the same time the police—and I—have the right to say to you, "You stupid idiot, what the hell were you thinking?"[42]

A recent study indicates that drunk women exhibit an "alcoholic myopia" that makes them more vulnerable to rape.[43] Most of us will not have been surprised by this confluence of science and common sense.

The Blank Slate: Pinker on the Arts

To Pinker, painting, music, and literature are pleasure technologies, manifestations of an evolutionary psychology that incorporates Stone Age preferences we cannot transcend. For example, the enjoyment of painting has its origin in the move from the forest to the savannah, an evolutionary shift that shaped our visual apparatus. This is why "a good landscape painting or photograph will simultaneously evoke an inviting environment and be composed of geometric shapes and pleasing balance and contrast."[44] While this observation may account for the pleasures of Claude Lorraine, it does nothing at all for Rembrandt, Jackson Pollock, or Marcel Duchamp. Rembrandt's arresting self-portrait in the Louvre is a testament to the ravages of old age and the depth of character it can bestow; Jackson Pollack's *Blue Poles*, in the National Gallery of Australia, is a testament to the daring of an artist who invented a technique in which the act of painting and its product perfectly coincide, a testament to the possibilities of spontaneous design. Marcel's Duchamp's *Fountain*, a urinal on exhibit, is a testament to the aesthetic possibilities of everyday objects, those that while not previously noticed as works of art are nonetheless masterpieces of design.

Pinker gives music and literature analogous treatment. Since babies prefer consonant musical intervals over dissonant ones, he feels adults must follow suit, preferring "Jingle Bells" to Janáček. For Pinker, some natural sounds—thunder, rushing water, birdsong—can affect us emotionally. It is because of this that we find a stripped-down version of these at the heart of our favorite melodies. Were this true, Pinker would have us enjoying Richard Strauss's *Don Quixote* because a flock of baa-ing sheep makes a cameo appearance, and Mahler's Sixth because cowbells are employed.[45] In the case of literature, Pinker asserts that we naturally prefer stories with a happy ending, the Heidi novels to *War and Peace, The Odd Couple* to *Othello*. Pinker's explanations of the arts suffer from his persistent confusion of the pleasant and the pleasurable.

It is this confusion that leads to his wholesale dismissal of modern and postmodern art. The poet and critic Randal Jarrell pinpoints the fallacy that Pinker's view embodies: "The general public . . . has set up a criterion of its own, one by which every form of contemporary art is condemned. The criterion is, in the case of music, melody; in the case of painting, representation; in the case of poetry, clarity."[46] Pinker further assumes that modern and postmodern works of art are unique in their radical departure from accepted norms, that their initially hostile or indifferent reception is an exception. But painters, composers, and writers have often extended the reach of their arts to the detriment of their popularity. No one at the time thought that Schubert's *Winterreise* was an immortal masterpiece; no one at the time thought Beethoven's last quartets were music.

Pinker's dismissal of sophisticated art parallels his dismissal of sophisticated literary criticism, talk about the arts that assumes that knowledge enhances pleasure. He excoriates Judith Butler because of her jargon-ridden prose:

> The move from a structuralist account in which capital is understood to structure social relations in relatively homologous ways to a view of hegemony in which power relations are subject to repetition, convergence and rearticulation brought the question of temporality into the thinking of structure, and marked a shift from a form of Althusserian theory that takes structural totalities as theoretical objects to one in which the insights into the contingent possibility of structure inaugurate a renewed conception of hegemony as bound up with the contingent sites and strategies of the rearticulation of power.[47]

But Pinker's prose can be equally opaque when he is doing what Butler exactly is doing—talking as an expert to experts

> The findings of Experiments 1–3 demonstrate that readers extract the semantic number of bare nouns automatically and represent it in a way that is comparable to the conceptual number that they extract from visual perception. Because these effects of number congruency were observed when lexical semantic features are absent (for nonwords, used in Experiment 3), these results demonstrate that semantic number can be extracted via grammatical knowledge from morphological marking alone. The grammatical computation of semantic number can result in Stroop-like interference or facilitation in the judgment of the number of words in the display, despite the irrelevance of the grammatical number of the word to the task demands.[48]

Pinker on Violence

In the Rwanda genocide of 1994, as many as one million died, approximately 10% of the population; in the Syrian civil war, 400,000 have died, nearly 2% of the population. In the United States, 10% of the population would mean 32 million deaths; 2%, over six million. In the face of atrocity so devastating, Pinker's claim in *The Better Angels of Our Nature: Why Violence Has Declined*[49] seems incredible. Yet his 95 graphs and tables tell the same story: violence has indeed declined. Asked in an interview why he felt he could write about history, presumably a discipline beyond his expertise, Pinker focused on these very graphs and tables: "I'm an experimental psychologist. *Better Angels* concentrates on quantitative history: studies based on datasets that allow one to plot a graph over a timeline. This involves the everyday statistical and methodological tools of social science, which I've used since I was an undergraduate—concepts such as sampling, distributions, time series, multiple regression, and distinguishing correlation from causation."[50]

In table 10.1 below, the death tolls for historical large-scale atrocities are given as two values: estimated toll from historical records and that same toll adjusted to the world's population in the mid-20th century.[51] The most startling number in this startling list is the 36 million deaths in the An Lushan Revolt during China's Tang Dynasty. Scaled up to account for population growth, we have the mid-20th-century equivalent of 429 million deaths, a staggering number greater than the entire populations of the United States and Canada, and far greater than the total death toll from both world wars. It would seem that Pinker has indeed discovered an astonishing pattern in human history, a legitimate source of the scientific sublime.

Or has he? Operational definition is surely among the methodological tools of the social science Pinker employs. But neither from his tables and figures nor from his text can an operational definition of violence be derived. Wars and preventable famines are both consequences of state power, but only the former involves violence. Nor is slavery primarily an exercise in violence, though human beings are enslaved as its consequence, and slavery leaves slaves open to violence exercised with impunity. Moreover, Pinker's own figures can be differently aggregated to tell another, less optimistic story. Any account of 20th-century violence must notice that two world wars, Mao's famine, Stalin's persecutions, the Congo war, and the Russian and Chinese civil wars accounted for 154 million deaths, all in only five decades. By contrast, the Mongol conquests (adjusted rank no. 2) resulted in 40 million deaths (278 million equivalent) in 163 years. Indeed, six of the top ten in Pinker's murderers' row of mass violence span more than a century. Moreover, the statistic of 36 million deaths in the nine years of the An Lushan Revolt is controversial, as Pinker admits. That total is based on a highly questionable census taken at the rebellion's end, with the reigning Tang Empire in disarray and greatly shrunk in size. A more conservative estimate is 13 million unadjusted.[52]

Added to these problems are Pinker's omissions. If Pinker sees slavery as a form of violence, now thankfully for the most part eliminated, why does he omit child labor and child soldiering, two forms of enslavement still with us? Why does he omit the deaths resulting from America's love affair with the private automobile, the state's decision to exercise its muscle in favor of unsafe over safe transportation, in effect to license a form of

TABLE 10.1 Historical Death Tolls Adjusted to World Population in the Mid-20th Century

UNADJUSTED RANK	CAUSE	CENTURY	DEATH TOLL	DEATH TOLL: MID-20TH-CENTURY EQUIVALENT	ADJUSTED RANK
4	An Lushan Revolt	8th	36,000,000	429,000,000	1
3	Mongol Conquests	13th	40,000,000	278,000,000	2
9	Mideast Slave Trade	7th–19th	19,000,000	132,000,000	3
5	Fall of the Ming Dynasty	17th	25,000,000	112,000,000	4
15	Fall of Rome	3rd–5th	8,000,000	105,000,000	5
11	Timur Lenk (Tamerlane)	14th–15th	17,000,000	100,000,000	6
7	Annihilation of the American Indians	15th–19th	20,000,000	92,000,000	7
10	Atlantic Slave Trade	15th–19th	18,000,000	83,000,000	8
1	Second World War	20th	55,000,000	55,000,000	9
6	Taiping Rebellion	19th	20,000,000	40,000,000	10
2	Mao Zedong (mostly government-caused famine)	20th	40,000,000	40,000,000	11
12	British India (mostly preventable famine)	19th	17,000,000	35,000,000	12
17	Thirty Years' War	17th	7,000,000	32,000,000	13
18	Russia's Time of Troubles	16th–17th	5,000,000	23,000,000	14
8	Josef Stalin	20th	20,000,000	20,000,000	15
13	First World War	20th	15,000,000	15,000,000	16
21	French Wars of Religion	16th	3,000,000	14,000,000	17
16	Congo Free State	19th–20th	8,000,000	12,000,000	18
19	Napoleonic Wars	19th	4,000,000	11,000,000	19
14	Russian Civil War	20th	9,000,000	9,000,000	20
20	Chinese Civil War	20th	3,000,000	3,000,000	21

violence? And what about our lax gun laws, unbudgeable even in the face of child massacre? These categories show that we may not be less violent, only differently violent, the victims of new expressions of violence shaped by shifting social, political, and economic circumstances.

The questionable aspects of Pinker's claim extend to his explanation. While in the last 200,000 years we have not evolved anatomically, we have, Pinker avers, evolved socially and politically: gradually, reason has triumphed over irrationality. Indeed, "once a society has a degree of civilization in place, it is reason that offers the greatest hope for further reducing violence."[53] The prospects of a reduction in violence seem particularly bright because we are getting smarter. Are we really? While it is true that we are getting better at IQ tests, a phenomenon known as the Flynn effect, we cannot infer from this that human intelligence is actually increasing. Flynn agrees: "Can anyone take seriously the notion that the generation born in 1937 was that much more intelligent than the generation born in 1907, to say nothing of the generation born in 1877?"[54] The Flynn effect is only about what intelligent tests measure.

In any case, getting smarter hardly means getting nicer. Take practical reason, our ability to deal more and more effectively with manufacturing problems. This is an area where we have undoubtedly improved over time. But the same assembly line that produced Model T Fords made the Holocaust possible. Nor is being smart necessarily correlated with being free from socially harmful bias. Martin Heidegger and Gottlob Frege are two philosophers with deservedly towering reputations: Frege was an anti-Semite; Heidegger, a Nazi. Ezra Pound and T. S. Eliot were poets and critics of great distinction: both were anti-Semites. Pound was a fascist and a traitor to boot.

Conclusion

A well-respected experimental psychologist, Steven Pinker burst upon the scene with his first book, *The Language Instinct*, a masterly performance that makes us feel that our unique ability to speak and understand language is a daily miracle linguistic science has completely unraveled. *The Blank Slate* argues that our genetic heritage must be factored into any explanation of our social behavior. *How the Mind Works* is another blockbuster; there Pinker presents us with a completely worked-out computational theory of our mental processes. *The Better Angels of Our Nature* reinterprets history, insisting that, contrary to popular belief, violence has diminished over time. Every one of Pinker's major works shares a single overriding assumption: science can be relied on to shed significant light on subjects far removed from the laboratory or the observatory; science can astonish us by its revelations about language, about the mind, about human behavior generally, and about violence in particular.

It takes nothing away from Pinker the polymath to say that his books consistently violate a rule that another scientist of note, Richard Feynman, followed in his popular science writing: never stray from knowledge that could legitimately appear in a textbook for graduates or undergraduates as well-established science. While Pinker's consistent violation of this principle in no way constitutes a criticism of his achievement—he has many interesting and provocative things to say on a wide variety of subjects—it does indicate that on these he should be read with caution. We must treat with a certain skepticism his view that a computer model can explain how the mind works, that ordinary readers can decide between rival theories of language generation and the origin of concepts, and that we live in a safer world because we are smarter than our ancestors. In each of these cases, we may well be prompted to ask whether the scientific sublime has been legitimately invoked and evoked.

CHAPTER 11 | Richard Dawkins

The Mathematical Sublime

What had that flower to do with being white,
The wayside blue and innocent heal-all?
What brought the kindred spider to that height,
Then steered the white moth thither in the night?
What but design of darkness to appall?—
If design govern in a thing so small.

—ROBERT FROST

IN AN EPISODE of *The Simpsons*, "Black Eyed, Please," Ned Flanders has a nightmare. He visits his "personal hell" where they "worship famous atheist Richard Dawkins, author of *The God Delusion*," a devilish figure in the process of "making Catholic-saint stew." Irreverent enough to be attracted to the program's irreverence, and enough of a celebrity to be asked to do the show's voice-over, Dawkins is content to appear as a parody of himself. But his skepticism is no act. It is deep-seated, with roots in his early childhood. Concerning his 18-month-old self, Dawkins says:

> At Christmas a man called Sam dressed up as Father Christmas and entertained a children's party in Mrs. Walter's house. He apparently fooled all the children, and finally took his departure amid much jovial waving and ho-ho-ho-ing. As soon as he left, I looked up and breezily remarked to general consternation, "Sam's gone!"[1]

This precocious skepticism blossoms in Dawkins's later views, a set of convictions in which science does not so much supplement as substitute for religion: "a friend . . . persuaded me of the full force of Darwin's brilliant idea and I shed my last vestige of theistic credulity probably about the age of sixteen."[2] To Dawkins, biology is no more—and no less—than a rigorous skepticism applied to the living world. No need for Father Christmas.

Without question, Dawkins's vision of biology, a living world ruled by mathematics, is a "grand conception,"[3] readily comparable to the origin stories of Weinberg, Greene, Randall, and Hawking, a saga of "how unordered atoms could group themselves into ever more complex patterns until they ended up manufacturing people."[4] In his work, Dawkins has employed mathematics to create, as Adam Smith said of Copernicus, "another constitution of things, more natural indeed, and such as the imagination can more easily attend to, but more new, more contrary to common opinion and expectation, than any of those appearances themselves."[5] For Dawkins and Smith alike, scientific discovery generates an authentic sense of the sublime: "What is new and singular, excites that sentiment which, in strict propriety, is called Wonder; what is unexpected, Surprise; and what is great or beautiful, Admiration."[6] To Dawkins, evolution not only shows how the living world develops; it shows how evolution itself evolves, and how culture evolves, an all-encompassing theory. Nevertheless, Dawkins and Smith part company on a key issue. For Smith the experience of the sublime opens a door to another sublime, a realm beyond the world of experience. From nature's design, Smith feels, we must infer a designer: "The idea of a universal mind, of a God of all, who originally formed the whole, and who governs the whole by general laws, directed to the conservation and prosperity of the whole."[7] For Dawkins, there is a design, but no designer.

The Limits of Popularization

Can life be subject to mathematical principles? Is there a surprising concurrence between abstract systems of numbers and natural systems in the living world? In *On the Origin of Species*, Darwin broached the possibility:

> When we look at the plants and bushes clothing an entangled bank, we are tempted to attribute their proportional numbers and kinds to what we call chance. But how false a view is this! Everyone has heard that when an American forest is cut down, a very different vegetation springs up; but it

has been observed that the trees now growing on the ancient Indian mounds in the Southern United States, display the same beautiful diversity and proportion of kinds as in the surrounding virgin forests. What a struggle between the several kinds of trees must here have gone on during long centuries, each annually scattering its seeds by the thousand; what war between insect and insect—between insects, snails, and other animals with birds and beasts of prey—all striving to increase, and all feeding on each other or on the trees or their seeds and seedlings, or on the other plants which first clothed the ground and thus checked the growth of the trees! Throw up a handful of feathers, and all must fall to the ground according to definite laws; but how simple is this problem compared to the action and reaction of the innumerable plants and animals which have determined, in the course of centuries, the proportional numbers and kinds of trees now growing on the old Indian ruins![8]

The world had to wait decades for the outcome Darwin had so confidently foresaw, the rise of the mathematical sublime in evolutionary biology. Only with the rediscovery of Mendel in 1900 and the formulation of the Hardy-Weinberg law in 1908 did the mathematics of biology become a serious pursuit. G. H. Hardy's case is an especially telling example of the power of mathematics to explain the living world. In a letter to *Science*, a mere mathematician was able to assert with supreme confidence that "there is not the slightest foundation for the idea that a dominant character should show a tendency to spread over a whole population, or that a recessive should tend to die out."[9] Brown eyes, however dominant, would never stamp out blue eyes, however recessive. This was a mathematical fact; incredibly, it was also a scientific fact. Employing mathematics in this manner, the biologist imitates the physicist. For both the central task is now "to correlate the incoherent body of crude fact confronting [them] with some definite and orderly scheme of abstract relations, the kind of scheme [they] can borrow only from mathematics."[10] Still, Darwin's vision was far from fully realized. Only in later decades was mathematics applied systematically to problems of animal behavior. In the twenties, the Lotka-Volterra predator-prey equations were discovered; in the sixties, R. H. MacArthur and E. O. Wilson gave us their insights into island biogeography; in the seventies and eighties, John Maynard Smith applied von Neumann's game theory to the interactions among living things.

While Dawkins follows in Maynard Smith's mathematical footsteps in his popularizations, it is out of the question for him to present the actual mathematics of the concepts he wishes to explain. Nevertheless, there is no reason for such general readers to fret. "The elegant ideas" of game theory," he asserts, "can be expressed in words without mathematical symbols, albeit at some cost in rigor."[11] In *The Selfish Gene*, we discover the degree to which this claim is true. Dawkins makes two attempts to define a central concept derived from game theory, the evolutionary stable strategy, or ESS. In the first edition, he calls it "as a strategy which, if most members of a population adopt it, cannot be bettered by an alternative strategy."[12] In the revised edition, thirty years later, he says that "an ESS is a strategy that does well against copies of itself."[13] The first definition does not tell us the conditions that must pertain for a strategy to be an ESS; the second gives us only one of the conditions. Dawkins admits that the definition is "incomplete."[14] A paper by Maynard Smith and G. A. Parker gives us the complete definition, necessarily mathematical:

I is an ESS if, for all alternative strategies J, either
$$E_I (I) > E_I (J) \tag{1a}$$

or

$$E_I (I) = E_I (J) \text{ and } E_J (I) > E_J (J) \tag{2b}^{15}$$

In other words, I is an evolutionary stable strategy if and only if the expected gain of employing it against species $_I$ is greater than the expected gain of employing strategy J. There is another possibility. I is an evolutionary stable strategy if the expected gain of employing it against species $_I$ is equal to the expected gain of employing strategy J and if, at the same time, the expected gain of employing I against species $_J$ is greater than the expected gain of employing J against species $_J$. If this is plain English, it is surely not plain enough. The only conclusion we can draw is that elegant mathematics does not translate easily into elegant prose.

It is to Dawkins's credit that he does not hide this difficulty from his readers. He conveys the essential message: words must inevitably shortchange the elegance mathematics routinely achieves. This translation is a task whose essence is encapsulated in Herbert Simon's famous coinage: satisficing, bypassing the good for the good enough. To popularize mathematics, Dawkins analogizes, always informing his readers of the limits of his comparisons. He discusses a game relevant to the ESS, the "war of attrition," a contest in which species threaten but never carry out their threats. Time is the only resource they expend, waiting for their rival's reaction. To explain this mathematical game, Dawkins employs the analogy of an auction sale: the resource the animal "is competing for may be valuable but it is not infinitely valuable. It is only worth so much time and, as at an auction sale, each individual is prepared to spend only so much on it."[16]

There is, however, an important difference between a war of attrition and an auction sale. In the latter, Dawkins makes clear, only one person walks away with the goods for which only he has paid. In contrast, in a war of attrition, although only one animal walks away with the goods, both pay the price in time and effort. Worse, if both animals bid up to the maximum, no one gets the goods. Given these possible outcomes, it would seem, the best strategy is to give up at the start, to look elsewhere to satisfy one's needs. But if such a strategy were employed, what would be the result? As more and more animals pulled out, waiting around a minute or two more would soon become the better strategy. The minutes it would be profitable to wait would go up and up until the maximum was reached, equal to the perceived value of the resource. At this point, it would make sense again to pull out at the beginning. There would be a continual oscillation.

Dawkins points out a problem with this perfectly sensible conclusion. It is mistaken: "By using words, we have talked ourselves into picturing an oscillation in a population. . . . Mathematical analysis shows that this is not correct."[17] In fact, there is an evolutionary stable strategy, one that only mathematical analysis reveals. It consists in being always unpredictable. Animals, however, can *be* unpredictable only if they *appear* unpredictable, only if they exhibit "a poker face."[18] So a poker face would evolve as an ESS. You might think that an alternative strategy would also evolve: lying. Not so. With the advent of lying, another strategy would evolve: calling the bluff. Lying is an unstable strategy. We see from this example, that while words must short-change mathematics, analogies can betray it. But not in Dawkins's hands, a way of writing that rewards the careful reader.

A mathematical theory of animal behavior, however elegant, is not a fact; it becomes a fact only when tested against the world and not found wanting. Dawkins does an excellent job of explaining this transition. Although in pioneering work Maynard Smith employed ESS to explain some aspects of animal behavior, only a field biologist could test his theory. Maynard Smith did his best; he mined the literature:

G. A. Parker (1970) has described a situation which agrees rather well with Eq. 2. Female dung flies of the genus *Scatophaga* lay their eggs in cow pats. The males stay close to the cow pats, mating with females as they arrive to lay their eggs. What strategy should a male adopt? Should he stay with a pat once he has found one, or should he move on in search of a fresh pat as soon as the first one begins to grow stale? . . . What is significant is that

Parker was able to show that the expected number of matings was the same for males which left early as for those which stayed on. This means that the males are adopting an ESS; natural selection has adjusted the probability of leaving per unit time to bring this about.[19]

Dawkins pinpoints the state of the art when Maynard Smith wrote this article: "The theory of evolutionary stable strategies was rich in cunning speculation and poor in hard data."[20]

Because he was also a theoretical biologist, Dawkins was in no better position than his mentor to test the ESS against the world. When he met Jane Brockmann, however, his prospects brightened. She had spent over 1,500 hours observing and recording the egg-laying behavior of the female *Sphex ichneumoneus* wasp. She found that sometimes these wasps dug their own burrows; these were the diggers. Sometimes they attempted to occupy burrows dug by other wasps; these were the enterers. Was there an ESS? If so, how could it be revealed?

Dawkins's and Brockmann's first assumption, that the proper unit with which to measure evolutionary success was the behavior of the individual wasp, got them nowhere. Then Alan Grafen entered the picture:

> when we were actually wrestling with the difficulties of wasp analysis, one of our main leaps forward occurred when, under the influence of A. Grafen, we kicked the habit of worrying about individual reproductive success and switched to an imaginary world where "digging" competed directly with "entering."[21]

The assumption behind the original idea was deeply mistaken. There were in fact no digging and entering wasps; there were instead wasps that sometimes "chose" to dig and sometimes to enter. What had to be compared was not the performance of individual wasps but two strategies for evolutionary survival that issued, eventually, in an ESS.

In *The Extended Phenotype*, what had been a demonstration of ESS becomes, more boldly and powerfully, a demonstration of the theoretical value of imagining a unit of selection other than the individual organism. To Dawkins, insect behavior should be viewed as if under the influence of a computer program with two subroutines, a striking analogy:

> The digging routine "runs" in a large number of different physical wasp nervous systems. Entering is the name of a rival subroutine which also runs in many different wasp nervous systems, including some of the same

physical nervous systems as, at other times, run the digging subroutine. Just as a particular . . . computer functions as the physical medium through which any of a variety of chess programs can act out their skills, so one individual wasp is the physical medium through which sometimes the digging subroutine, at other times the entering subroutine, acts out its characteristic behavior.[22]

The success of this research strategy—of treating digging and entering as if they were units of selection—should not mislead us into thinking that they *are* units of selection. The ESS that emerges, the durable balance between digging and entering, is determined, ultimately, by natural selection at the level of genes.

The Selfish Gene and *The Extended Phenotype* do not simply see evolution from the point of view of the gene rather than the individual plant or animal; they see evolution from points of view other than the individual and give those points of view a mathematical realization. Without leaving his professional or his amateur readers behind, Dawkins explains this significant theoretical shift. The remarkable achievement of both books is emphasized in a review by Maynard Smith:

> I have left till last what is to me the strangest feature of [*The Selfish Gene* and *The Extended Phenotype*], because I suspect it will not seem strange to many others. It is that neither book contains a single line of mathematics, and yet I have no difficulty in following them, and as far as I can detect they contain no logical errors. Further, Dawkins has not first worked out his ideas mathematically and then converted them to prose; he apparently thinks in prose, although it may be significant that, while writing *The Selfish Gene*, he was recovering from a severe addiction to computer programming, an activity which obliges one to think clearly and to say exactly what one means. It is unfortunate that most people who write about the relation between genetics and evolution without the intellectual prop of mathematics are either incomprehensible or wrong, and not infrequently both. Dawkins is a happy exception to this rule.[23]

Are Genes Really Selfish?

Dawkins's polemical skills are extraordinary, so extraordinary that they sometimes lead him and his readers astray, surprising them, it is true, but dismaying them as well. An instance is his insistence on ideologically laden terms like the *selfish* gene. Of course, he reminds us that he doesn't really mean *selfish*. While it is true that no scientist would take this terminological problem seriously—after all, do quarks have flavors?—*The Selfish Gene* is not meant for scientists only. Philosopher Mary Midgley puts her finger on the problem for general readers, the sense that they are no longer in charge of their lives:

> The message which any unsophisticated reader must receive is that he is a helpless pawn in the hands of his physical inheritance and can therefore stop trying, because he has a universal excuse. Or—to look at it another way—any states of discouragement and depression which may overwhelm him are veridical. What he feels when in their grip is the truth. He really is not capable of acting, and the sense of agency which he has at other times is a delusion. Paralysis of the will is his real condition. Since people sometimes believe the books they read, this seems a message which one should think twice about printing. [24]

This dire prediction is realized by at least one reviewer on the Amazon website, a response Dawkins quotes only to say that it is "preposterously mistaken."[25] I reproduce a section of the review he omits:

> The book renders a God or supreme power of any sort quite superfluous for the purpose of accounting for the way the world is, and the way life is. It accounts for the nature of life, and for human nature, only too well, whereas most religions or spiritual outlooks raise problems that have to be got around. It presents an appallingly pessimistic view of human nature, and makes life seem utterly pointless; yet I cannot present any arguments to refute its point of view. I still try to have some kind of spiritual outlook, but it is definitely battered, and I have not yet overcome the effects of this book on me.

To Dawkins, our genes use us as vehicles to carry out their own purposes. But are we really just "vehicles," subject to the extreme passivity that the term evokes? Philosopher David Hull objects to Dawkins's coinage for this reason. He suggests that vehicles be rechristened interactors, because "a terminology that tends to exaggerate the causal role of replicators at the expense of the different but equally important causal role of what I term interactors is less than perspicuous."[26] Hull's criticism is much closer to home than Midgley's; it is friendly criticism. While she has no use for *The Selfish Gene*, Hull's praise is lavish: "*The Selfish Gene* was published in 1976. Dawkins wrote the book to educate general readers about biological evolution, to prod experts out of their slumbers over the issue of individual selection, and to catch the imagination of students. He succeeded on all counts."[27] Well, perhaps not on *all* counts.

The selfish gene is just one Dawkins analogy among many that might lead the unwary to conclude that human beings are simply complex mechanisms. Dawkins also compares us to computers and robots; he personifies genes as oarsmen, comparing the cooperative efforts of a complex of genes to the analogous efforts of eight oarsmen working together to win the Oxford-Cambridge boat race. While he makes it clear that this is only an analogy, on the same page he says unequivocally, "The oarsmen are genes."[28] It gets worse. He does not merely compare us to robots; he identifies us as robots created and controlled by our "replicators," our genes:

> Now they swarm in huge colonies, safe inside lumbering robots, sealed off from the outside world, communicating with it by tortuous indirect routes, manipulating it by remote control. They are in you and in me; they created us, body and mind, and their preservation is the ultimate rationale for our existence.[29]

Dawkins never disavows the mechanistic implications of this passage. In an addendum to the anniversary edition, published thirty years later, he tells us that robots "are capable of intelligence, learning and creativity." He asks: "What on earth do you think you are if not a robot, albeit a very complicated one?[30]

There is a problem in equating robots and us. Consciousness is, arguably, a state that we share with no machine. Dawkins assumes that computer chess programs act *as if* they were conscious; in fact, he applauds a journalist's use of "the language of human consciousness" when describing the activities of the chess program Deep Thought. Deep Thought, the journalist writes, "contemptuously brushes aside" Igor Ivanov's "desperate fling."[31] Consciousness is involved in this exchange, of course, but it is the consciousness of the human beings who created Deep Thought. This indelible fact makes it difficult to agree with Dawkins when he asserts that a chess computer that wins a world championship constitutes "a lesson in humility."[32] In effect, he forgets what he has already said: "The evolution of the capacity to simulate seems to have culminated in subjective consciousness. Why this would have happened is, to me, the most profound mystery facing modern biology."[33]

This ideological bias should not mislead us into undervaluing the contribution of this marvelous book. *The Selfish Gene*—and Dawkins's work as a whole—is characterized not by one but by two central themes. The first grounds our altruism in our genetic selfishness. This tells us why parents care for their offspring: they share so many more genes than cousins or strangers. The more genes that are shared, the greater the potential for altruistic behavior. The problem is with the scope of Dawkins's explanation. It cannot explain why parents abuse their children; it cannot explain why in a 2010 Afghanistan firefight William "Kyle" Carpenter fell on a grenade to shield a fellow Marine from the blast.[34] There is another theme, however, one that is in no way ideological: the contention that the genes are the proper units of selection. This is a bold. novel, and well-defended vision of evolutionary theory:

> Each entity must exist in the form of lots of copies and at least some of the entities must be *potentially* capable of surviving—in the form of copies— for a significant period of evolutionary time. Small genetic units have these properties: individuals, groups, and species do not. It was the great achievement of Gregor Mendel to show that hereditary units can be treated in practice as indivisible and independent particles. Nowadays we know that this is a little too simple. Even a cistron [DNA section coding for a protein subunit] is occasionally divisible and any two genes on the same chromosome are not wholly independent. What I have done is to define a gene as a unit which, to a high degree, *approaches* the ideal of indivisible particularateness.[35]

There is a problem with this formulation, a problem that Dawkins recognizes and solves. It has been objected that the notion of a gene for saving someone from drowning makes no sense. But "all you have to concede is that it is possible for a single gene, other things being equal and lots of other essential genes and environmental factors being present, to make a body more likely to save somebody from drowning than its allele [its opposite number] would."[36]

Evolvability

To explore the possibility that evolution itself evolves, that there exists a new permutation in a theory that itself exemplifies the scientific sublime, Dawkins creates a computer program that selects for two traits, symmetry and segmentation. Biological symmetry and segmentation are, of course, generated by DNA according to a developmental plan internal to the embryo, one that Dawkins's computer program simulates. In figure 11.1, the program

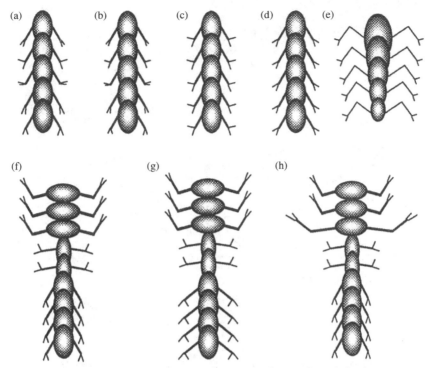

FIGURE 11.1 Computer generated "insects" with kaleidoscopic variation in segmentation.

SOURCE: Richard Dawkins, *Climbing Mount Improbable* (New York: Norton, 1996), 246. Copyright © by Richard Dawkins. Original drawings copyright (©) 1996 by Lalla Ward. Used by permission of W. W. Norton &Company, Inc.

generates a range of virtual insects that exhibit symmetry and segmentation, segments that vary in size and shape and in leg angle and length.

In figure 11.2, we see that real animals also exhibit these transformations. Dawkins explains:

> The world is populated by major groups of animals—arthropods, molluscs, echinoderms, vertebrates—each one of which has a form of kaleidoscopically restricted embryology which has proved evolutionarily fruitful. Kaleidoscopic embryologies have what it takes to inherit the earth. Whenever a major shift in kaleidoscopic mode or "mirror" has spawned a successful evolutionary radiation, that new mirror or mode will be inherited by all the lineages in that radiation. This is not ordinary Darwinian selection but it is a kind of high-level analogy of Darwinian selection. It is not too fanciful to suggest as its consequence that there has been an evolution of improved evolvability.[37]

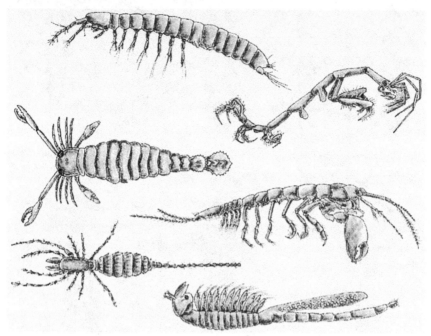

FIGURE 11.2 Segmentation variation in real animals.

SOURCE: Richard Dawkins, *Climbing Mount Improbable* (New York: Norton, 1996), 254. Copyright © by Richard Dawkins. Original drawings copyright (©) 1996 by Lalla Ward. Used by permission of W. W. Norton &Company, Inc.

Although Dawkins writes a computer program that strongly suggests evolvability, it does not confirm it; theories do not come with the evidence for their truth attached. Dawkins's instincts, however, are on the right track. There is real evidence that evolvability is a fact of nature. Peter K. Dearden and Michael Akam show that the fertilized eggs of grasshoppers increase in segmentation and symmetry according to a genetic plan.[38] What is the link between such research and evolvability? Symmetry and segmentation are instances of compartmentalization, the channeling of development as we move from genes to characters to functions. Figure 11.3 shows that evolutionary change is facilitated by compartmentalization into master traits such as symmetry and segmentation "making various cell populations independent, reducing the chance of lethal mutation and increasing the independence of variation and selection within a compartment."[39] Although there are many causal arrows from G1, G2, and G3 to C1 and from G4, G5, and G6 to C2, there are only three between C1 and C2, a significant paucity whose outward signs in the case of grasshoppers are symmetry and segmentation.

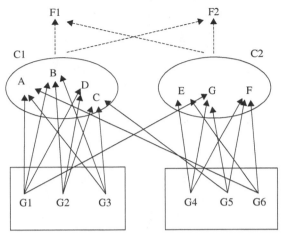

FIGURE 11.3 A model for segmentation and symmetry from genes to characters, each with its own function.

SOURCE: Gunter P. Wagner and Lee Altenberg, "Complex Adaptations and the Evolution of Evolvability," *Evolution* 50 (1996): 971.

And not only in the case of grasshoppers. Our arms are shorter than our legs, a characteristic we share with our first cousins, the great apes, but not with our second cousins, the monkeys. We walk and run from place to place; so do the great apes. Monkeys, on the other hand, swing through trees. Somewhere along the evolutionary line, then, the constraints on limb equality were eased for us and for our first cousins. We became more evolvable in one direction than the monkeys. We can see the results of this trend in figure 11.4.[40] Somewhere along the line, the development of our arms and legs parted company, another example of compartmentalization, of evolvability evolving.

Cultural Change

What if the scope of evolutionary theory included not only us but everything we create: our art, our music, the cars we drive, the way we shop, the games we play? What if, in addition to genes that shape our nature, there exist memes that shape our culture? Think of the game of bridge as a meme with two modules: bidding and playing. There are rules for each; there are also strategies for bidding and playing according to these rules. Like genes, memes are replicators, and the human beings who apply them are their vehicles. At our local club, the bridge meme goes to work, a meme that may also be realized physically: we use plastic boards to insure

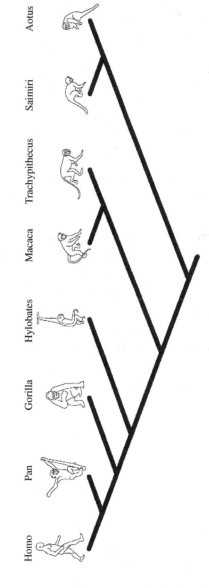

FIGURE 11.4 Human and apes part company with monkeys.

SOURCE: Nathan M. Young, Günter P. Wagner, and Benedikt Hallgrímsson, "Development and the Evolvability of Human Limbs," *Proceedings of the National Academy of Science, USA* 207 (2010): 3402.

that rivals in competitive bridge play the exact same hands, maximizing skill over chance. The bridge meme itself has evolved: the once popular game of whist was overtaken by auction bridge. Contract bridge eventually arose, and auction bridge went extinct.

In his belief in memes, in cultural evolution, Dawkins is not alone. Donald Campbell, E. O. Wilson, and the team of Robert Boyd and Peter Richerson were prominent pioneers in cultural evolution. Wilson's *Sociobiology* was an intellectual blockbuster; Boyd and Richerson's collaboration was equally impressive, leading to three books with telltale titles: *Culture and the Evolutionary Process, The Origin and Evolution of Cultures*, and *Not by Genes Alone*. In 2006, there appeared in an issue of *Behavioral and Brain Sciences* "Toward a Unified Science of Cultural Evolution," a review essay whose message was clear. Despite a welter of theories, despite a lack of consensus among them, Dawkins, Campbell, Wilson, and Boyd and Richerson were responsible for a conceptual tide not likely soon to subside.[41]

This persistence does not mean that the idea of cultural evolution is free from criticism. Philosopher Mary Midgley quotes Dawkins, to his apparent detriment:

> When we look at the evolution of cultural traits, and their survival value, we must be clear whose survival we are talking about. . . . A cultural trait may have evolved in the way that it has, simply because it is *advantageous to itself*. . . . Once the genes have provided their survival machines with brains that are capable of rapid imitation, the memes will automatically take over. We do not even have to posit a genetic advantage in imitation.[42]

Midgley comments acidly: "So apparently if we want to study (say) dances, we should stop asking what dances do for people, and should ask only what they do for themselves."[43] Clearly this is not what Dawkins has in mind. Instances of cultural survivals advantageous only to themselves are not hard to find: "mission creep" in Vietnam, Afghanistan, and Iraq; nuclear proliferation; and the bombing of German civilian targets during World War II culminating in the firestorm in Dresden.

We might also criticize meme theory because Dawkins's own examples are unimpressive: "tunes, ideas, catch-phrases, clothes fashions, ways of making pots or of building arches."[44] A better example would be the scientific article, a genre that originated in 17th-century England and France, a form of prose that had by the 21st spread to the whole of science. Such articles exhibit three heritable modules: style, structure, and argument. Over time, their style evolves from literary to plain. Their structure tightens, evolving from the essayistic to the functional, eventually generating a master finding system of titles, keywords, headings and citations, equations and figures. Argument—the core of scientific articles—evolves from informal to formal, a tightly controlled interaction of words and images.[45]

Do animals other than ourselves possess memes? Do songbirds have a culture? Writing half a century ago, Dawkins could find only one songbird study that supported this claim. Subsequent literature reveals successful work on chickadees,[46] dark-eyed juncos,[47] cowbirds,[48] and silvereyes.[49] In Cardoso and Atwell, for example, two populations of juncos were studied: city dwellers and their country cousins. While the songs of these two populations can differ markedly, when Cardoso and Atwell measured the difference between their two populations, it turned out that cultural mutations—meme modifications—accounted for 56% of the difference. They concluded that

> in this species, memes can be replaced not only by cultural selection or drift, but also by de novo creation of memes (learning new song types from improvisation . . .), which is another form of cultural mutation. Because many urban song types were sung by a single male, new song types appear to contribute to maintain a diverse meme pool and, thus, should have influenced the pattern of meme replacement.[50]

We think it is beyond doubt that Dawkins's memes have contributed to cultural evolution. Was this what he meant to do? Not according to Dawkins. Of his introduction of memes in the final chapter of the first edition of *The Selfish Gene*, he says: "My primary intention . . . was not to make a contribution to the theory of human culture, but to downplay the gene as the only conceivable replicator that might lie at the root of a Darwinian process."[51]

Is Design Possible without a Designer?

According to evolutionary theory, creatures as well-designed as cheetahs and sharks must be the consequence not of planning but merely of selection operating on random variation and mutation, selection whose products engage in a war of all against all. Only the fittest survive: the cheetah with the most powerful legs, the shark with the sharpest teeth. To demonstrate the plausibility of this explanation, Dawkins introduces a computer program. He writes this program with only two modules, *repro-duce*, to imitate the evolutionary march from generation to generation, and *develop*, to imitate individual development, the march from eggs to children. In his program

> all of these children are mutant children of the same parent, differing from
> their parent with respect to one gene each. This very high mutation rate is a
> distinctly unbiological feature of the computer model. In real life, the prob-
> ability that a gene will mutate is often less than one in a million. The reason
> for building a high mutation rate into the model is that the whole perfor-
> mance on the computer screen is for the benefit of human eyes, and humans
> haven't the patience of a million generations for a mutation![52]

Figure 11.5 illustrates a typical result. A virtual insect develops from a real dot on the computer screen. While this transformation is a far cry from real evolution triggered by natural selection—there is no environment the "insect" inhabits and no other creatures with which it interacts—nevertheless the case for the dramatic effect of gradual change is well made: "Jumping could *theoretically* get you the prize faster—in a single hop. But because of the astronomical odds against success, a series of small steps, each one building on the accumulated success of previous steps, is the only feasible way."[53]

In *Climbing Mount Improbable*, a decade later, Dawkins creates a computer program that operates more nearly like natural selection. He imitates the struggle of existence by recreating the perennial contest between the spider and the fly. There are many ways that animals can narrow the odds between them and their prey—the claws of the eagle, the tongue of the chameleon—and the spider's web. Webs are architectural triumphs whose structural brilliance we routinely ignore. Finely knit to entangle prey, they have just enough adhesive strength to capture, yet they are not so finely knit or so adhesive that the spider itself is tangled in its own snare. In figure 11.6 are displayed some results of the gradual development Dawkins has built into his program. The most successful web with these characteristics will not be the one that captures the most flies but the one that captures the most with the least expenditure of the spider's resources. It is not the one on the bottom right; it is the one in the bottom middle. The program is another victory for gradual change as an engine of evolution, illustrated in a simulation more realistic than the first. At the same time, it is an assault on deity; in more and more realistic simulations, Dawkins has generated a design without a designer.

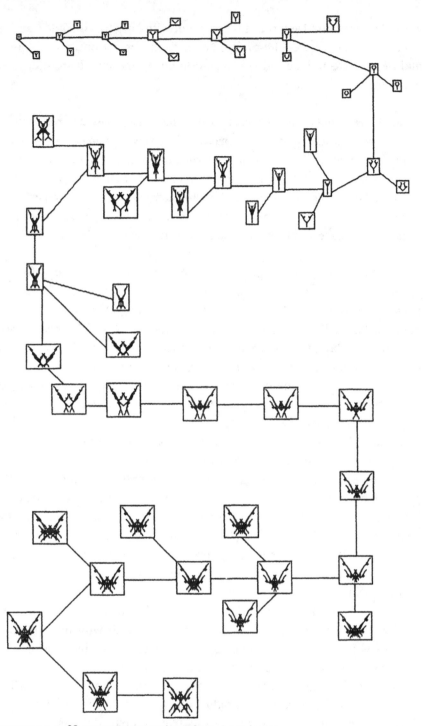

FIGURE 11.5 29 computer generations, from dot to insect.

SOURCE: Richard Dawkins, *The Blind Watchmaker: Why the Evidence of Evolution Reveals a Universe Without Design* (New York: Norton, 1996). Copyright © 1996, 1987, 1986 by Richard Dawkins. Used by permission of W.W. Norton & Company, Inc.

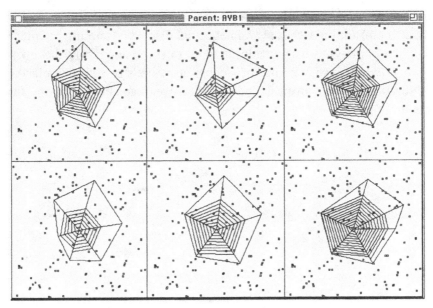

FIGURE 11.6 Computer generated spider webs.

SOURCE: Richard Dawkins, *Climbing Mount Improbable* (New York: Norton, 1996), 61. Copyright © by Richard Dawkins. Original drawings copyright (©) 1996 by Lalla Ward. Used by permission of W. W. Norton &Company, Inc.

Conclusion

During the trial of the knave of hearts, Alice rudely interrupts the queen, who insists that the sentence must precede the verdict. "Stuff and nonsense," Alice says. "Off with her head," responds the queen. But her ukase is to no avail. Now grown to her full height—a presage that her dream of Wonderland is about to end—Alice responds dismissively: "You're nothing but a pack of cards." As if by magic, figures of authority turn back into colored cardboard. In her defiance of authority, it is her innocence that saves Alice from the charge of arrogance, a defense on which Dawkins cannot rely. What has he to say to readers of *The Selfish Gene* who are in no position to balance the claims to evolutionary centrality of the genes, the individual, and the group? What is one to say who cannot reliably sift metaphor from fact?

I do not think his brilliance as an expositor of mathematically oriented biology can absolve Dawkins from the charge of irresponsibility implicit in these questions. In the end, however, it is his brilliance that saves him from opprobrium. To read him, right or wrong, is never less than to engage in a pleasurable joust with a passionate and interesting man, never less than to embark on a voyage of intellectual discovery into the biological sublime powered by mathematics, a series of surprising results that evoke on nearly every page a sense of wonder at nature's secrets and a lasting admiration for mathematical modeling that reveals and explains them. Just not all of them.

CHAPTER 12 | E. O. Wilson
The Biophilic Sublime

More is a man of an angel's wit and singular learning. I know not
his fellow. For where is the man of that gentleness, lowliness and
affability? And, as time requireth, a man of marvelous mirth and
pastimes, and sometime of as sad gravity. A man for all seasons.

—ROBERT WHITTINGTON

Whilst seated on a tree, & eating my luncheon in the sublime solitude
of the forest, the pleasure I experience is unspeakable.—The number
of undescribed animals I have taken is very great. . . . I attempt class
after class of animals, so that before very long I shall have notion
of all.—so that if I gain no other end I shall never want an object of
employment & amusement for the rest of my life.[1]

—CHARLES DARWIN

SITTING IN THE same rain forest where Darwin penned these words more
than a century earlier, E. O. Wilson shares the identical "cathedral feeling," the identical sense of the biological sublime evoked by the diversity
of the biosphere: "Hold[ing] still for long intervals to study a few centimeters of tree trunk or ground, [and] finding some new organism at each shift
in focus." [2] It is a feeling for his fellow creatures exhibited in every aspect
of E. O. Wilson's life: his efforts to understand ant society, his discovery
of sociobiology as means of understanding all societies, and, finally, his
efforts to preserve the diversity of the biosphere in which all societies must
find their place. For Wilson, environmental ethics flows naturally from
the cathedral feeling he shares with Darwin, their sense of the biological
sublime.

E. O. Wilson, Scientist

It is 1969 and there is a knock on Wilson's office door. It signals the arrival a talented German entomologist, Bert Hölldobler, invited to Harvard for an extended stay. The two hit it off almost immediately; eventually, Hölldobler returns to Harvard as a full professor. While his command of English improves, his German accent never entirely disappears, an accent, Wilson feels, that lends weight to his lectures: his is the authentic voice of German science. Friends and collaborators, the two produce a stream of science that culminates in the publication of *The Ants*. The award of the Pulitzer for this book is Wilson's opportunity simultaneously to signal and make light of his achievements. The book "weighed 7.5 pounds, fulfilling my criterion of a magnum opus—a book when dropped from a three-story building is big enough to kill a man."[3] The award is also an opportunity to express his gratitude to the institution that continued to nurture his talent. At the monthly college meeting, he recalls, "I stood and basked in the applause of the Harvard faculty. Bless my soul, the Harvard faculty. Where could I go from here but down."[4]

By this time, however, his co-author has left Harvard. The daily drudgery of turning out grant proposals has taken its toll. Although all his grants were funded, Hölldobler's resources fell short of his soaring entomological ambitions. In an instant, however, his problems are solved. He gets a call from his alma mater, the University of Würzburg, offering him the leadership of the newly formed Theodor Boveri Institute of Biological Science. While his friendship with Wilson endures, the move marks the end of their close collaboration. Hölldobler does not leave America entirely behind; he and his family spend their summers in Arizona, where he studies American ants.

Wilson's collaboration with Hölldobler is based on a synergy of the two research traditions. Wilson is a population biologist; Hölldobler is an ethologist. The object of study of population biology is the nest; that of ethology, the behavior of its inhabitants. Wilson is a synthesizer; Hölldobler anchors Wilson's broader vision in the stubborn facts of insect behavior, the apparently insignificant details of their daily lives. These "little things" can sometimes matter:

> In the course of our work, Bert's attention fastened on fragments of old cocoon silk plastered on the walls of the *Prionopelta* nests. He asked, as much to himself as to me, What does this mean? Nothing, just trash, I answered. When the new adults emerge from the cocoons, their nestmates

throw out the silk fragments, and they don't bother to stack them in separate garbage dumps. No, no, he said, look: the pieces are lined up as a smooth layer on the gallery walls.[5]

Pursuing this insight, Hölldobler discovers that these fragments were converted to wallpaper designed to control the climate of the nest. Wilson's deference to Hölldobler in this case is not false modesty. He is the senior colleague, far more famous than his friend. But the first author of *The Ants* is Hölldobler.

Richard Lewontin is a distinguished biologist who summarily dismisses the sociobiology Wilson considers his crowning achievement. He calls Wilson's defense, *Genes, Mind, and Culture*, "absurd." Characterizing Wilson as "a past master of the art of rhetorical inflation," he says that sheer rhetoric "would seem to be his major contribution to the book."[6] Understandably, Wilson begs to differ. He feels that Lewontin's attack on sociobiology is ideological, that his science has been hijacked by his Marxism, his conviction that "only anti-reductionist non-bourgeois science would help humanity attain the ultimate, highest goal, a socialist world."[7]

But Wilson neither belittles nor scorns his most formidable opponent. In fact, Lewontin's work at the University of Chicago is so impressive that Harvard considers inviting him to join its faculty. While some professors fear he would be a divisive force, Wilson advocates for the appointment. A senior colleague, George Kistiakowsky, calls unbidden to tell him that he is making a grave mistake. Perhaps Wilson should have listened. *Sociobiology* is reviewed in the *New York Review of Books*; soon thereafter, the *Review* publishes a letter from Science for the People, a blistering polemic against the book. In his reply, Wilson expresses his dismay and hurt feelings. He has been ambushed by colleagues. He points out correctly that their letter twists his words out of their proper context, that it makes a careful scientist look like a tawdry charlatan.

Worse is to come. At the annual meeting of the American Association for the Advancement of Science, one consequence of this attack becomes evident. Wilson is on the speakers' platform when a group of students chant, "Racist Wilson, you can't hide, We charge you with genocide." Soon thereafter, a student pours a pitcher of water over Wilson's head as the demonstrators shout, "Wilson, you're all wet." Stephen Jay Gould (yes, *the* Stephen Jay Gould), a member of Science for the People, takes the microphone to deplore the attack.[8] But it is too late. The damage has been done: Science for the People has turned an important scientist into a cardboard villain. Even after these untoward events, however, Wilson's high opinion of his formidable opponent remains unchanged. Richard Lewontin, he says,

> was the kind of adversary most to be cherished, in retrospect, after time has drained away emotion to leave the hard inner matrix of intellect. Brilliant, passionate, and complex, he was stage-cast for the role of contrarian. He possessed a deep ambivalence that kept friend and foe off balance: intimate in outward manner, private inside; aggressive and demanding constant attention, but keenly sensitive, anxious to humble and to please his listeners at the same time, intimidating, yet easily set back on his heels by a strong response, revealing a fleeting angry confusion that made one—almost— wish to console him.[9]

His portraits of his friend and collaborator and of his most formidable scientific antagonist exemplify Wilson's ability to convey in a few pages the kind of man and the kind of scientist Wilson is.[10]

Wilson and the Ants

Wilson relates that for the renowned apiologist Karl von Frisch, bees were a magic well: the more he drew from it, the more there was to draw.[11] With von Frisch as their model, Wilson and Hölldobler devote their professional lives to the "elegant analysis of complex systems"[12] of ant social interaction. Their book on the leafcutter ants is an example, a survey of one member of a remarkable segment of this insect population, the eusocial, ants who cooperate so fully that we can almost say that they function as a single organism. Reading this book, we see two myrmecologists at their no-nonsense best, generating from the simple recital of facts an authentic sense of the biological sublime:

> One typical *Atta sexdens* nest, more than six years old, contained 1,920 chambers, of which 238 were occupied by fungus gardens and ants. The loose soil that had been brought out and piled on the ground by the ants during the excavation of their nest weighed approximately 40,000 kilograms (40 tons).[13]

Construction operated all the more swiftly because the ants carried loads four or five times their body weight, the equivalent of our carrying twelve

FIGURE 12.1 Out for a stroll.

times the weight of the heaviest barbells sanctioned by the International Weightlifting Federation. Figure 12.1 shows that seeing is believing.

Wilson and Hölldobler not only want us to read about the construction of these nests; they also want us to see them. The cutaway diagram in figure 12.2 shows both the exterior and interior structure; the man in the background reveals the size. We can view this illustration with confidence because other intrepid entomologists poured cement or molten metal into some of these vast nests; then, after the cement had hardened or the metal cooled, they excavated them, taking at every step the same care as archaeologists, careful not to damage what their extraordinary exertions have revealed.

In *The Journey of the Ants*, a popular collaboration also written with his friend and collaborator Hölldobler, the bare facts themselves incite wonder. The nests of one species consist of pavilions constructed from leaves. To bring together leaves too far apart for a single insect to join, the ants form a living chain:

> The lead worker seizes a leaf edge with her mandibles and holds fast. The next worker then climbs down her body, grips her waist, and holds on. A third worker now climbs down to grip the second worker's waist, and so on ant upon ant, until chains ten workers or more are formed, often swinging free in the wind. When an ant at the end of the chain finally reaches the edge of a distant leaf, she fastens her mandibles on to it, closing the span of the living bridge, and all the entrained force begins to haul back in an attempt to bring the two leaves together.[14]

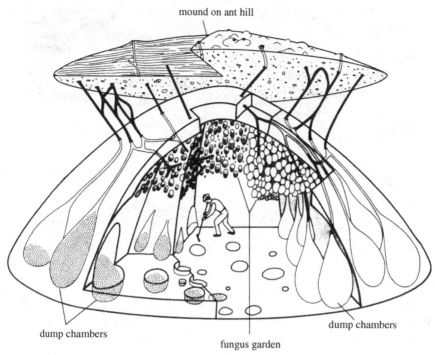

mound on ant hill

dump chambers

dump chambers

fungus garden

FIGURE 12.2 The nest of a leafcutter ant colony. The dump chambers store the waste that remains after the fungus has consumed the nutritious elements in the leaves the ants transport to the nest.

SOURCE: Bert Hölldobler and E. O. Wilson, *Journey to the Ants: A Story of Scientific Exploration* (Cambridge, MA: Harvard University Press, 1994), 116.

For the products of such innate collective behavior, Richard Dawkins has coined a felicitous and accurate phrase: "extended phenotype." Just as the ants' physical characteristics are the product of natural selection, so are their nests. They are "an exemplary illustration of natural selection acting at the level of the entire society."[15]

In *Journey of the Ants*, we join Wilson, following a living river from its source to its mouth:

Wilson was spellbound by the sight of one of foraging expeditions of *Atta cephalotes*. At dusk on the first day in camp, as the light failed to the point where he and his companions found it difficult to distinguish small objects on the ground, the first worker ants came scurrying purposefully out of the surrounding forest. They were brick red in color, about 6 millimeters in length, and bristling with short, sharp spines. Within minutes several hundred had entered the campsite clearing and formed two irregular files that passed on either side of the biologists' shelter. They ran in nearly straight lines across the clearing, their paired antennae scanning right and left, as though drawn by some directional beam on the far side of the clearing. Within an hour, the trickle expanded to twin rivers of tens of thousands of ants running ten or more abreast.[16]

Wilson and Hölldobler have created a bond with their readers, giving them the vicarious experience of the natural sublime.

Leafcutter ants farm: they grow edible fungus that feeds on leaves the ants bring back to the nest. For their harvesting journeys, they create an elaborate network of trails, such as those shown in figure 12.3. This particular trail network encompasses about 100,000 square feet. Given an ant's one-millimeter stride, traversal from the nest to the most remote of leafcutter locations involves about 200,000 steps—in human terms, about ten miles. On their return, moreover, the ants carry a leaf fragment many times their body weight, one on which a smaller sister may well be perched to ward off parasitic flies.

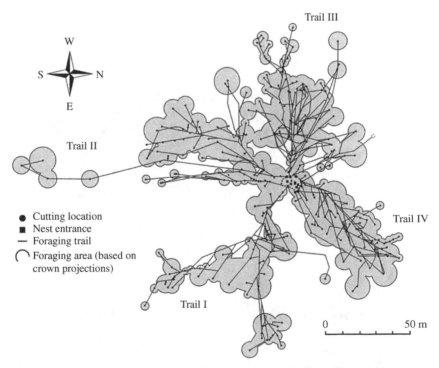

FIGURE 12.3 Foraging trails of the leafcutter ant *Atta colombica*. The nest is dead center.

SOURCE: Bert Hölldobler and Edward O. Wilson, *The Leafcutter Ants: Civilization by Instinct* (New York: W. W. Norton, 2010), 125.

Each of these fragments must be harvested. First, an ant must mount a leaf; then, using her mandibles, the equivalent of jaws, she must cut into the leaf, as a chain saw cuts into timber. Wilson and Hölldobler's description of the process exemplifies one of the myrmecologist's chief tools: incredible patience. Days of trained observation are condensed into thirty seconds of reading:

> During the opening of the motile mandible, the cutting jaw is pushed against the leaf by lateral head movements. Next, the motile mandible is closed, pulling the cutting jaw further against the leaf and lengthening the incision. In this phase, the motile mandible also moves deeper into the leaf surface, thus preparing the way for the cutting jaw. As soon as both jaws meet, the cycle starts again. Thus, one jaw functions as "cutting knife" and the other one as "pacemaker."[17]

Figure 12.4 that illustrates this process illustrates another as well, stridulation, the rapid repeated motion indicated by a two-headed arrow, a movement that facilitates the incision.

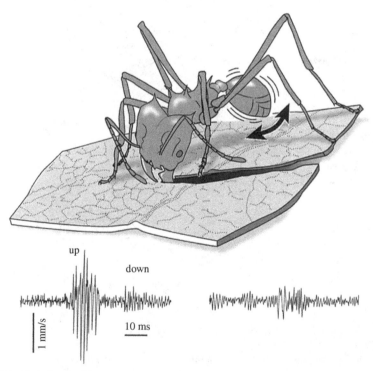

FIGURE 12.4 Leafcutting.

SOURCE: Bert Hölldobler and Edward O. Wilson, *The Leafcutter Ants: Civilization by Instinct* (New York: W. W. Norton, 2010), 64.

After harvesting, the leaves are returned the nest. In figure 12.5, the narrative form of the comic strip is harnessed to convey the process by which these are turned into the substrate for the leafcutters' fungus gardens.

First, the leaves are transferred to smaller workers who clip them into fragments. Then even smaller workers turn these into moist pellets seasoned with fecal droplets. Next, even smaller workers insert the pellets into a larger mass of similar material. Then the fungus is planted on this new substrate. Finally, other workers tend the garden.

The stridulations that steady the activity of the leafcutter ants serve another purpose. They signal other ants nearby that a new, rich source of vegetation has been discovered. Stridulation, however, is just a minor kind of formic communication. The main form is chemical. Chemicals leave trails to tell their sister ants where their prey is to be found or where danger lurks. Chemicals, like those illustrated in figure 12.6, are the ants' eyes and ears.

Ants have a chemical "vocabulary" of 10 to 20 phrases, enabling them to express attraction, to recruit, to convey alarm, to assist in the identification of other castes and of larvae, and to discriminate between nestmates and strangers. In his novel *Anthill*, to bring us closer this world, Wilson leaves scientific prose behind: "The olfactory world . . . contained much more than an invisible roadmap. Bombarding the ant from below and from all sides above were the odors of organisms that inhabited the soil—so densely as to make up a large part of the physical bulk of the soil."[18]

Although ants are eusocial, this central characteristic of their behavior excludes neither struggles for dominance within the nest nor warfare between communities. Within the nest, the ants, all sisters, turn sibling rivalry into a routine dominance-submission ritual, a form of behavior not unheard of in human families. Ant colonies, however, are far more aggressive than humans in their relationship with others. In warfare, their strategy and tactics seem to betray the study of Clausewitz's *Vom Kriege*. They scout; they set up perimeters; they bluff by blowing themselves up to appear larger than they are; they drop pebbles on the enemy, a unique instance of tool use.

In telling the story of these ardent warriors, Wilson and Hölldobler perform at an Olympic level. They deploy analogy: ant colonies at war are "like so many Italian city-states." After giving us a blow-by-blow account of combat between deadly enemies, the *Pheidol* and the fire ants, they tell us that even in defeat, the *Pheidol* soldiers remain steadfast and unyielding, a stance that reminds the authors of another, long-ago battle:

The *Pheidol* soldiers remain true to their caste. They do what they are programmed to do: stay and fight to the death. They are the insect equivalents of the Spartan defenders who held fast and died at Thermopylae before the Persian hordes, to be commemorated by a metal plate inscribed, "Stranger, if you see the Lacedaemonians, tell them that we lie here faithful to our instructions."[19]

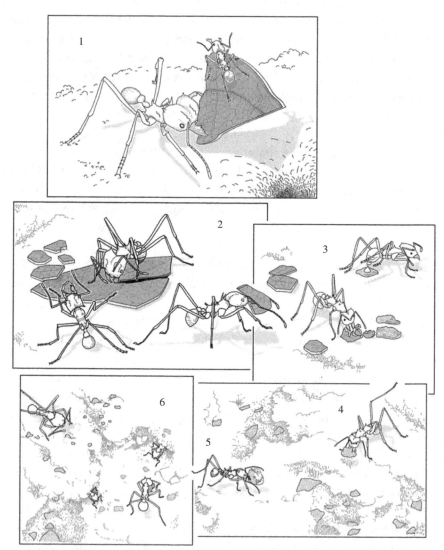

FIGURE 12.5 Leaves are turned into substrate for a fungus garden.

SOURCE: Bert Hölldobler and Edward O. Wilson, *The Leafcutter Ants: Civilization by Instinct* (New York: W. W. Norton, 2010), 54.

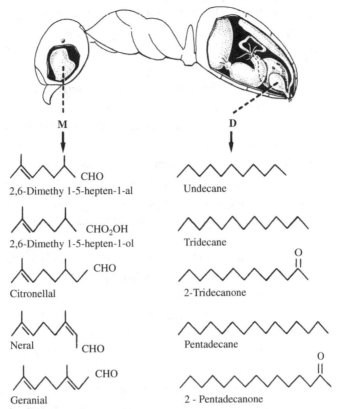

FIGURE 12.6 The mandibular gland (*M*) and the Dufour's gland (*D*).

SOURCE: Bert Hölldobler and E. O. Wilson, *Journey to the Ants: A Story of Scientific Exploration* (Cambridge, MA: Harvard University Press, 1994), 56.

Readers need not experience these battles at secondhand. Magnifying glass in hand, they can closely observe "a bare patch of lawn on which hundreds or thousands of pavement ants are engaged in mortal combat."[20]

Our sense that the world of ants participates in the sublime stems from our surprise that creatures with brains no larger than a grain of sand possess the equivalent of a culture. It is this clue, this sense that a complex culture is written in the genes of even these allegedly simple creatures, that leads to Wilson's most arresting theory, sociobiology, the idea that every animal culture, including ours, is written in the genes.

Sociobiology

When Wilson published *Sociobiology*, Science for the People denounced the book unequivocally. It was not just another permutation of evolutionary theory; it was a repellant polemic in favor of a specious biological determinism that supported the status quo, a society patently unfair to women and minorities:

> Wilson dissociates himself from earlier biological determinists by accusing them of employing an "advocacy method" (deliberately selecting facts to support preconceived notions) generating unfalsifiable hypotheses. He purports to take a more solidly scientific approach using a wealth of new information. We think that this information has little relevance to human behavior, and the supposedly objective, scientific approach in reality conceals political assumptions. Thus, we are presented with yet another defense of the status quo as an inevitable consequence of "human nature."[21]

In fact, Wilson has no hidden political agenda; that he does is a Marxist fantasy. Not so easily dismissed, however, is the charge that "this information has little relevance to human behavior." In both *Sociobiology* and in *Consilience*, a book that followed a quarter century later, the scope of sociobiology remains the same: evolutionary theory explains not only nature but culture as well. This contention seems to stretch evolutionary theory past the breaking point. For example, Wilson contends that music is a trait we share with whales and birds, one that evolved in each species as an instrument of social cohesion.[22] Although there is no reason to doubt this conclusion, nevertheless it does not follow that "even great works of art might be understood fundamentally with knowledge of the biologically evolved epigenetic rules that guided them."[23] It does not follow because something extraordinary happened to music in the West, a cultural change that biology cannot plausibly explain.

Beginning in the Middle Ages, music became a sophisticated cultural pursuit. A complex notational system evolved, a set of instructions designed to preserve compositions, a form of cultural memory. While these notations constrain and shape performance, they do not determine it: "In all traditions, including our own, notation under-determines performance and identifies works of music only when read in the context of a performance tradition."[24] This underdetermination means that "a person who performs woodenly, unfeelingly or grotesquely may make all the sounds prescribed by the score, but he will show nevertheless that he has understood nothing. Musical understanding is a form of aesthetic understanding: it is manifest in the conscious search for the right phrasing, the right dynamics, the right tempo."[25] Consider William James's amusing aside: "A Beethoven string-quartet is truly, as someone once said, a scraping of horse's tails on cat's bowels, and may be exhaustively described in such terms, but the application of this description in no way precludes the simultaneous applicability of an entirely different description."[26]

For the listener of music of the highest order, of Beethoven's piano sonatas, for example, musical understanding is a form of connoisseurship. Reviewer William Youngren says of Malcolm Bilson's recording of Beethoven's *Tempest* sonata that it

> is in all ways excellent: for the vigor and extension of line in the first movement; for the dignified, fastidious slow movement; and for the (once again) highly improvisatory finale, in which the beat keeps getting lost—just as Beethoven intended. This sonata makes the listener attend more closely to the sound and the capabilities of Bilson's instrument. Many of the rapid left-hand figures have an extraordinary clarity, and the arpeggios in the slow movement are dazzling.[27]

No one thinks that these words—that any words, in fact—describe the message the music conveys to the listener. To pianist and conductor Daniel Barenboim, "it must be understood that one cannot explain the nature or the message of music through words. . . . When we try to describe music with words, all we can do is articulate our reactions to it, and not grasp music itself."[28] Youngren is neither describing nor explaining the music; he is describing and explaining his reaction to it; he teaching us how to listen. Youngren is to music what Eric Asimov is to wine.

There is another issue: Bilson plays the sonata, not on a modern piano but on its ancestor, the fortepiano. Of another fortepiano performance of this sonata, reviewer Peter Burwasser says: "To miss hearing Beethoven played on period instruments, especially at the level of excellence we get with [Ronald] Brautigam, is to miss something special and even revelatory."[29] It seems plausible to say that we should hear the sonata as Beethoven's audience might have heard it, Bilson and Brautigam's aim. It seems equally plausible to say that since Beethoven himself saw the *Tempest* as a departure from his previous practice, he envisioned his sonata played on an instrument with capabilities that did not yet exist, the modern concert grand.

Regardless of the instrument employed, there is a difference of opinion over how the sonata realizes its composer's desire that "from today on I shall take a new path."[30] In my view, musicologist Janet Schmalfeldt gets the heart of the matter:

> I argue that Beethoven's music initiates new stylistic directions whereby, as conventional Classical formal processes become gradually transformed or

"deformed," new cases of genuine formal ambiguity increasingly arise. In such cases . . . it is as if the composer invites the performer to play a deter-minative role in our understanding of the formal process. Thus the alliance between composer and performer—with both of these understood as listen-ers par excellence, to say the least—grows all the stronger as composers of the early nineteenth century respond to the impact of Beethoven's music.[31]

In this paragraph, we see connoisseurship at work in aid of historical recon-struction. About such matters, evolutionary theory must remain silent.

Neither can evolutionary theory tell us anything interesting about ethi-cal decision-making. For that we need philosophers. In *On Human Nature*, Wilson asks: "Can the cultural evolution of higher ethical values gain direction and momentum on its own and completely replace genetic evo-lution? I think not. The genes hold culture on a leash. The leash is very long, but inevitably values will be constrained in accordance with their effect on the human gene pool."[32] Philosopher Bernard Williams begs to differ. Because science takes the position that we do not occupy a special place in the natural world, it can have nothing interesting to say about eth-ics: "The aim of ethical thought is to help us to construct a world that will be our world, one in which we have a social, cultural, and personal life."[33]

How do ethical dilemmas manifest themselves? Let us suppose you are a ship's officer on the *Titanic*, just sunk in icy North Atlantic waters. You are manning a full lifeboat. Around you, struggling to board are desper-ate passengers flailing in icy waters. To prevent them from boarding, you beat them away with an oar. You are only doing your duty as a ship's offi-cer; you are saving what lives you can. That your own life is also saved is incidental. You have done nothing wrong; your behavior is in no way malicious; you do what you do in anguish. When you return safely, you are haunted by your action. A friend tries to comfort you, saying, "Don't give it a second thought; clearly, you did the right thing. Feeling regret is just irra-tional." Surely, this can never be the thing to say. In Williams's words, "We must rather admit that an admirable moral agent is one who on occasion is irrational. This is a new position; it may well be correct."[34] This same analysis applies to a truck driver who accidently runs over a child. If he is a moral agent, he is bound to feel regret, while the passenger sitting beside him is not so bound.[35] In these cases, ethics and rationality part company.

In another of Williams's thought experiments, we meet a fictional Gauguin, an aspiring painter who decides to spend the rest of his life pur-suing his art. He moves to Tahiti, leaving behind his large family to fend for themselves. In judging his conduct, does it matter whether through lack of talent or lack of application he fails as a painter? It does, for then we will say he was wrong to abandon his family. If, on the other hand, he succeeds in creating an artistic legacy all the world admires, we might very well judge that, while it is unfortunate that he left his family, he was not wrong to pursue a life that allowed his talent to flourish. The accident of success alters our judgement; our intuition that moral judgments always supersede other judgments has been undermined. Still,

> to regard Gauguin's decision simply as a welcome incursion of the amoral is anyway too limited. It will be adequate only if he is the amoral Gauguin. . . . If he is not, then he is himself open to regrets for what he has done to others, and, if he fails, then those regrets are not only all that he has, but . . . he no longer even has the perspective within which something else could be laid against them. That can make a difference to the moral spectator. While he may admire the amoral Gauguin's achievements, and indeed, admire him, this other Gauguin is someone who shares the same world of moral concerns. The risks these agents run is a risk within morality, a risk which amoral versions of these agents would not run at all.[36]

We might not want to say of irrational regret in the first case or of morality's limits in the second that they are the consequences, even remote consequences, of our genes holding culture on a leash, however long.

We might think that while Wilson is overreaching in the case of music and ethics, he is surely on target when we come to more mundane matters like being born, a process every animal experiences; surely all behavior surrounding birth is genetically determined; surely in this case we are just another animal. But Wilson is mistaken even here. Ants and orangutans are born into social groups wholly determined by evolutionary forces. These are species-specific; they are the same whether the ants and orangutans are in the wild or in a zoo in Chicago. Newly born orangutans and ants just are; it does not make sense to say their birth also has symbolic significance. But human birth *always* has symbolic significance; it *always* means. Moreover, what it means is deeply influenced by the immediately surrounding culture. To be born male or female, for example, is not just a genetic accident; it is a fate, one that is significantly different in America and Saudi Arabia. Evolutionary theory can have nothing to say about these matters. What Bernard Williams says about the possible contribution of sociobiology to morality applies to cultural matters across the board: "It might be able to suggest that certain institutions or patterns of behavior are not realistic options for human societies. That would be an important achievement, but first sociobiology will have to be able to read the historical record of human culture much better than it does now."[37]

Sociobiology already has. There are two puzzling forms of altruism that are exclusively human. First, we identify with far larger groups than do animals, with other Americans, for example, or other Catholics. Maciej Chudek and Joseph Henrich propose an evolutionary account for the binding power of these larger social units, one based on norms:

> Norms create selection pressures for learners to act as though they live in a world governed by social rules that they need to acquire, many of which are prosocial. Young children show motivations to conform in front of peers . . . spontaneously infer the existence of social rules by observing them just once, react negatively to deviations by others to a rule they learned from just one observation, spontaneously sanction norm violators . . . and selectively learn norms (that they later enforce) from older . . . and more reliable . . . informants.[38]

There is a second altruism puzzle: we sometimes condemn, even punish, actions we regard as immoral, even when the perpetrators are relatives or friends, well integrated into our social group. Peter DeScioli and Robert Kurzban address this issue. In their model, conscience is a defense mechanism designed to avoid condemnation of our own actions. We condemn others for violations of the ethical precepts for which we would condemn ourselves.[39]

These sociobiological investigations have analogous aims. They are not about judging the quality of ethical decisions; they are about the social and psychological processes that have evolved that make ethical judgments possible. Both programs of research conform to Williams's observation about limits of sociobiology. Nevertheless, they do not undermine sociobiology's insight into ethics, any more than the views of musicians, musicologists, and music critics undermine sociobiology's insight into music as an instrument of social cohesion. Rather, they certify to Wilson's brilliance as a theorist whose reach has on occasion exceeded his grasp, a commonplace occupational hazard in any serious intellectual pursuit.

Defending the Biosphere

The ant family is a member of the order Hymenoptera, the class Insecta, the phylum Arthropoda, the kingdom Animalia, and the empire Eukaryota.

It is an empire that, along with the empire Prokaryota, constitutes the whole of the biosphere, in other words, all life on earth. We are all members of an evolving community with 1.4 million known species. These are as large as whales and as small as bacteria; as loved as pandas and as hated as bedbugs. Like orangutans, they survive in the heat of the tropics; like penguins, they flourish in the chill of Antarctica. Like cacti, they make their home in the parching desert; like acorn worms, they live comfortably on the ocean floor under pressures of three hundred atmospheres; like the bacterium *Pyrolobus fumarii*, they inhabit volcanic hydrothermal vents, flourishing at a temperature of nine degrees above the boiling point of water.[40] To find life, we can look up at the birds and the communities of creatures that inhabit the rainforest canopy, or down to the worms and ant colonies just beneath our feet. Moreover,

> you do not have to visit distant places, or even rise from your seat, to experience the luxuriance of biodiversity. You yourself are a rainforest of a kind. There is a good chance that tiny spider-like mites built nests at the base of your eyelashes. Fungal spores and hyphae on your toenails await the right conditions to sprout a Lilliputian forest. The vast majority of cells in your body are not your own; they belong to bacteria and other microorganismic species. More than four hundred such microbial species make their home in your mouth.[41]

Wilson concludes that "the most wonderful mystery of life may well be the means by which it created so much diversity from so little physical matter."[42]

The biosphere is not static; its equilibrium is in constant flux. Over geological time there have been several mass extinctions, the most famous involving the dinosaurs. Changes in the biosphere occur as a consequence of natural disasters, such as volcanic eruptions, or as a result of naturally changing conditions, such as the position of the tectonic plates and the state of the climate. When these winnowings are over, however, life always returns. On August 27, 1883, a volcano erupted on the South Pacific island of Krakatau. As a result, the center of the island was replaced by an undersea crater over four miles long and nearly nine hundred feet deep. The eruption also created an island, Rakata, approximately nine miles square and devoid of life. Life soon returned. In May, 1884, a French expedition discovered "one microscopic spider—and only one; this strange pioneer of renovation was busy spinning its web." On a small sister island that later emerged to the north, one that was also completely lifeless, an expedition a half century later discovered "72 species of spiders, springtails, crickets, earwigs, barklice, hemipterous bugs, moths, flies, beetles, and wasps."[43]

Krakatau was a natural experiment demonstrating that life is stubborn; it returns, even to barren islands. Krakatau was not, however, a controlled experiment, one from which scientific conclusions can legitimately be drawn. Wilson accepted the challenge of creating such an experiment, a test of the distance effect. After the repopulation of an island denuded of life, an island like Krakatau, he conjectured, the population would reach a dynamic equilibrium at a level that depended on its distance from other land masses. The more distant these were, the lower the level of repopulation:

> In the early 1960s I spent a great deal of time poring over maps of the United States, day-dreaming, searching for little islands that might be visited often and somehow manipulated to test the models of island biogeography. I thought a lot about insects, creatures small enough to maintain large populations in tight places. A full-blown bird or mammal fauna might require an island the size of Guernsey or Martha's Vineyard.[44]

The answer was to miniaturize. Four tiny islands in the Florida Keys were selected, each different in size and all at different distances from each other and from a larger land mass. A survey of insect life having been conducted, a pesticide company was hired to fumigate, creating "four miniature Krakatau's."[45] In less than a year, the distance effect was confirmed. The island closest to a large island had had 43 species and now had 44; a distant island started with 25 species and was back to 22.

The trouble nature causes nature repairs. When we are the cause, nature cannot keep up with our depredations. While on her own nature extinguishes perhaps one species per million a year, we have increased the annual pace to between one in a thousand and one in ten thousand. Sometimes depredation is the consequence of our inadvertence. In the early 1900s a giant ornamental snail was introduced to islands in the Pacific and the Indian Oceans. It did not behave well; it gobbled up crops and other snails. The solution decided upon was a form of biological control. A new snail was introduced for whom the giant snail would be prey. These newcomers did not behave well either; they went after the native snails, wiping out half the Hawaiian species and one-quarter of the species on Mauritius.[46]

Inadvertence can also lead to ecological collapse. The biosphere is an intricate network we are just beginning to understand; the consequences of disturbing it cannot always be anticipated. By the end of the 19th century, the sea otter had been hunted to near extinction. In a place where extinction was total, a surprise was in store. With no otters, their food supply, sea urchins, proliferated, reducing the amount of kelp and seaweed that is the sea urchins' chief food. With the kelp and seaweed gone, large stretches of the ocean floor turned into a virtual desert, a sea urchin barrens.[47]

These episodes pale before the deliberate destruction of the rainforests of Central and South America. These contain more species in less space than anywhere on earth. When we visit them, they seem robust: "Walk the floor of a tropical rain forest, searching for specimens of almost any group, whether orchids, or frogs, or butterflies. You will find that the species change subtly every hundred or thousand meters."[48] Although the rainforest looks robust, in fact it is fragile. Small scale farming, logging, and cattle ranching are the culprits that turn forests into deserts. Small-scale farms are created when vegetation is cut down and burned, a form of agriculture that destroys the already nutrient-poor soil. Cattle ranching and logging collaborate in this destruction, harvesting timber by clear-cutting and felling whole forests to create pasturage. These activities are eliminating nearly 50,000 square miles of forest cover a year, 1% of the total, an area as large as the state of Pennsylvania.

In pursuit of answers to the exact character of this depredation, Thomas Lovejoy and his team turned a substantial part of the Amazon rainforest into a giant, long-term experiment, one still ongoing after more than thirty years. Its goal: to test the resilience of the rain forest. In *Diversity*, Wilson reports the results a dozen years in; I can report on the results 32 years in. The effort is truly extraordinary: it involves convincing the Brazilian government and landowners to set aside parcels of rainforest of differing size, keeping together a team over an extended period, and obtaining funding for a long-term project. Figure 12.7 shows the location and extent of this remarkable experiment. Figure 12.8 shows one effect of clearing the rainforest, the high Bowen ratio that robs the soil of moisture, turning the forest into a desert.

Soil depravation is not the only problem confronting the rainforest. Thirty-two years of observation have shown that bees, wasps, flies, butterflies, and understory birds decline. They show that predatory and parasitic species are more vulnerable than others, a loss that significantly alters food webs. They show that hyperdynamism is also operative, creating, among other things, an overabundance of butterflies. On the whole, forest biomass diminishes and at forest edges is increasingly vulnerable to fire. Clear-cutting also affects hydrology and biochemical recycling. There is a decline in carbon storage capacity as well, a diminution that accelerates the pace of global warming.

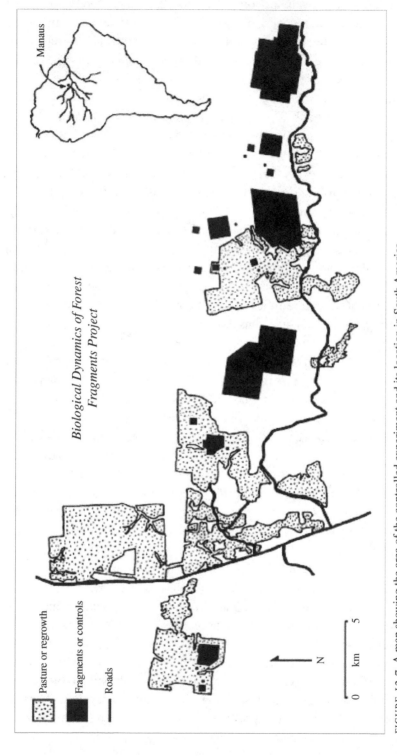

FIGURE 12.7 A map showing the area of the controlled experiment and its location in South America.

SOURCE: William F. Laurance et al., "The Fate of Amazonian Forest Fragments: A 32-Year Investigation," *Biological Conservation* 144 (2011): 58.

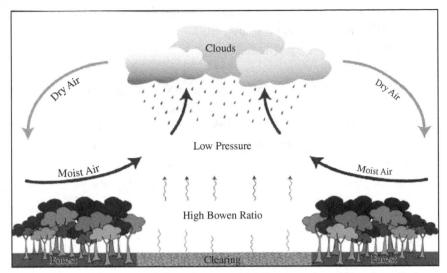

FIGURE 12.8 Clearing the rainforest creates a high Bowen ratio, robbing the soil of moisture.

SOURCE: William F. Laurance et al., "The Fate of Amazonian Forest Fragments: A 32-Year Investigation," *Biological Conservation* 144 (2011): 59.

Far from improving as a result of national policy, the situation is getting worse. A new highway is being built, and colonization by poor farmers is being encouraged. Lovejoy's 2011 review ends with a despairing passage, an emotional outburst that editors and referees permitted to stand:

> It is an uphill battle against a government bureaucracy that appears myopically determined to push ahead with colonization at any cost—despite the fact that colonists can barely eke out a living on the region's infamously poor soils. That such a globally important research project and conservation area could be lost seems unthinkable. That it could be lost for such a limited gain seems tragic.[49]

Although this review expressed pessimism about its future, the project, so far as I know, is still going strong.

Although the worldwide juggernaut against biodiversity seems unstoppable, Wilson remains optimistic. It is an optimism fueled by a universal human trait he feels he has identified, biophilia, an innate love for all living things, a genetic inheritance born not long after *Homo sapiens* first roamed the savannah. On its face, biophilia may seem like wishful thinking. In his contribution to the Wilson-edited *Biophilia Hypothesis*, Jared Diamond tells us that the native New Guinea tribes he studied displayed no affection whatever for their fellow creatures. Even with stone weapons, they hunted species to extinction; with better weapons, they simply got better at wiping species out.[50] Worse, tribesmen caused needless suffering in the animals they hunted:

> At the camp I found several New Guineans holding a large fruit bat of a species whose long and slender wingbones are used as nose decorations. The men wanted to cut these bones out of the bat's wing, but they did not bother to kill the bat before doing so. Instead, the two men spread out the bat's wings, another tied up the bat's mouth with a vine to prevent it from biting, and the other man then proceeded to dissect and scrape through the joints and muscles of the live bat so as to extract the bones.[51]

In contrast to Diamond, Wilson's co-editor, Stephen R. Kellert, rises to the defense of biophilia as "an ethical responsibility for conserving nature [that] stems . . . from more than an altruistic sympathy or compassionate concern: it is driven by a profound sense of self-interest and biological imperative."[52] Yet Kellert's own research shows that neither Americans nor Japanese exhibit this imperative. They are concerned only with those of their fellow creatures that have "prominent aesthetic, cultural, or historic features." They have no qualms about exploiting nature for practical purposes, even at the cost of significant environmental damage.[53]

Properly interpreted, however, biophilia is not nonsense at all: instead, it is a theme that has animated the whole of Wilson's professional life, a passion for biodiversity that leads inevitably to a passion for preservation. As a boy in Alabama, he read an article in *National Geographic* in which the entomologist William Mann rhapsodizes:

> Often have I gotten as big a thrill from a successful hunt for a rare ant as I have from the capture of giraffes or wart hogs. There is about as much physical exertion involved, too, turning over thousands of stones and logs, digging into the earth, chopping hard wood, and peeling bark from innumerable dead trees.[54]

Mann became Wilson's role model, his "hero from afar."[55] In his collection of essays *Biophilia*, the boy and the man are at one, providing testimony to an emotional arc stretching over a half century, a fusion of the mathematical and the dynamic sublime:

> A single bacterium possesses about ten million bits of genetic information, a fungus one billion, and an insect one to ten billion according to the

species. If the information in just one insect—say an ant or beetle—were to be translated into a code of English words and printed in letters of standard size, the string would stretch over a thousand miles. [This] lump of earth contains information that would just about fill all fifteen editions of the *Encyclopaedia Britannica*.[56]

Conclusion

In *The Theory of Island Biogeography*, E. O. Wilson and his co-author, Robert MacArthur, make a major contribution to theoretical biology; in *Sociobiology*, Wilson founds a new discipline; his Pulitzer-prize winning *The Ants* is a major contribution to entomology. But Wilson is not just a scientist of distinction; he is also a passionate advocate of environmental ethics. *The Diversity of Life* is a masterly plea for conservation; *Biophilia*, a collection of deeply moving essays that embody Wilson's life-long love affair with nature. Wilson is a novelist as well: *The Anthill* contains "The Anthill Chronicles," an impressive instance of the heroic epic. *The Naturalist* is his successful foray into autobiography. Wilson is also a distinguished pedagogue, responsible for the *Encyclopedia of Life*, a participatory website designed to catalog all life on earth. More recently, he launched and co-authored a revolutionary 3D high school biology textbook, *Life on Earth*, free through his Biodiversity Foundation. Last but not least, Wilson is a philosopher. *On Human Nature* and *The Meaning of Human Existence* argue powerfully in favor of secular humanism; *Consilience* argues powerfully in favor the unification of knowledge under strictly biological auspices. Central to all of these ventures is Wilson's biophilia, his deep affection for a biosphere that evokes in him a feeling for its sublimity and for the sublimity of the science that explains its workings, a feeling, he fervently hopes, he can instill in others. It is fitting that the subject of the final scientist-centered chapter of this book is E. O. Wilson, a man for all seasons.

PART III | Move Over, God

CHAPTER 13 | Move Over, God

Levin knew his brother and the way he thought; he knew that his
lack of faith had not come about because it was easier for him to live
without faith, but because the modern scientific explanations of the
phenomena of the universe had step by step squeezed out his beliefs.

—LEO TOLSTOY, *Anna Karenina*[1]

IN THE LAST two hundred years, physics, biology, and linguistics have
markedly increased our understanding of three mysteries that continue to
pique human curiosity: How did the universe begin? What is the origin of
life? What accounts for language, a capacity we share with no other creature? For helping us answer these questions, we have to thank linguists
like Steven Pinker, evolutionary biologists like Richard Dawkins, and theoretical physicists like Steven Weinberg. They each have a gift for translating important technical arguments in their respective disciplines into the
ordinary English we all can understand. Accompanying this extraordinary
achievement in popularization, however, there is an odd fellow traveler, a
campaign against religion, against God, a concerted attack extraordinary
in its persistence and its vehemence. In this final chapter, I would like to
investigate the extent, source, and nature of this attack, this insistence on
the part of so many scientists that they are not agnostics, properly skeptical of God's existence, but atheists, firm believers in his nonexistence, that
science, not God, is the only wellspring of the sublime.

It is the firmness of their belief that is in question. When Wolfgang Pauli
purportedly said of another scientist's work that "it is not right; it is not
even wrong," he meant that this work violated the boundaries of the discipline Pauli so successfully inhabited, that the offending scientist was
deluded when he thought he was doing physics. Of course, such professional judgments are far from perfect. Two eminent English mathematicians

had already dismissed the work of Ramanujan when G. H. Hardy, having received it unsolicited in the morning mail, judged its author as on a par with Euler or Gauss.[2] Still, this reversal of fortune is the exception, not the rule. When, however, even world-renowned scientists cross the border into a neighboring discipline in the humanities, say theology or biblical study, the exception is the rule: they dismiss what they do not trouble to understand. Their high intelligence and history of intellectual achievement tempts them to the conclusion that they need not, like any good anthropologist, inhabit for a time an alien world of discourse, the product of an intellectual nation whose citizens have devoted their lives to addressing two questions no science can successfully address: What is the meaning of life? What constitutes the good life? These scientists are the products of a myth according to which there is a general intelligence, g, a capacity that they possess in abundance, one that does yeoman's service in any area of human endeavor on which they choose to focus. So it is that Edward O. Wilson can write a book modestly titled *The Meaning of Human Existence*.

Particle physicists in particular are convinced that they view the world from an intellectual Everest only they have scaled; other scientists inhabit lower slopes; those in the humanities cluster forever at the mountain's base, enveloped in mist.[3] Eminent figures in the humanities—past luminaries—do not deserve their exalted reputations. Recalling his undergraduate days at Cornell, Steven Weinberg says, "My courses in philosophy left me puzzled about how ideas of Plato and Descartes that seemed to me absurd could have been so influential."[4] As a consequence of this attitude, our science popularizers misunderstand the attempts of theologians to offer proofs God's existence, attempts misconstrued as failed mathematical proofs; they misread the Bible, viewed as a chronicle of preposterous events ruled over by a nasty and arbitrary God; they misconstrue religion, seen as by its very nature hostile to science. While arrogance can explain their unwillingness to understand a continuing conversation about the meaning of existence and of the good life, a conversation whose inception is roughly coincident with our emergence as *Homo sapiens*, no trait of character can explain the vehemence of their attack on religion, their desire to make the natural sublime, not a manifestation of the divine but an end in itself. In his longitudinal study of the professional lives of physicists, Joseph C. Hermanowicz sees this attitude reflected even in the titles of popular books these scientists write:

Examining the titles [of books of popular physics], I note several patterns: a connection between physics and religion and/or spirituality; the idea of a search for deep answers about fundamental life questions; the notion that the answers are measurable through rational design; an anticipation that such answers would bring an "order" as yet unknown and unrealized in a modern and fragmented world; a recognition that such pursuit, and those engaged in it, brush with "greatness" as if to be god-like; and finally, the ultimate aim, or perhaps hope, that such revelation might entail a kind of immortality—or, at minimum, that the search may yield an ultimate permanence of understanding, of oneself, the world, and one's place in it, all based on *science*.[5]

Does God Exist?

It is not difficult to find evidence of the disbelief in God on the part of eminent popularizers of science. Steven Weinberg is certain that a final theory, a theory of everything, will not include a caring and concerned deity:

> Will we find an interested God in the final laws of nature? There seems something almost absurd in asking this question, not only because we do not yet know the final laws, but much more because it is difficult even to imagine being in the possession of ultimate principles that do not need any explanation in terms of deeper principles. But premature as the question may be, it is hardly possible not to wonder whether we will find any answer to our deeper questions, any sign of the workings of an interested God, in a final theory. I think that we will not.[6]

To E. O. Wilson, evolution is the source of the delusion theists share:

> The brain was made for religion and religion for the human brain. In every second of a believer's conscious life religious belief plays multiple, mostly nurturing roles. All the followers are unified into a vastly extended family, a metaphorical band of brothers and sisters, reliable, obedient to one supreme law, and guaranteed immortality as the benefit of membership.[7]

But it is Richard Dawkins in *The God Delusion* who trains his armamentarium solely on the Deity and his Word, the Bible. About the arguments for God's existence, the most famous of which is Anselm's, Dawkins is sardonically dismissive. Anselm asserts first that we can easily imagine a perfect being who exists only in our minds. If such an omniscient, omnipotent, and omnipresent being actually did exist, however, he would certainly be more perfect than this imaginary counterpart. Yet we can without doubt conceive of such a being. What is in our minds, therefore, cannot be *just* in our minds. Anselm also asserts that we can conceive of a being whose existence is necessary, who cannot be conceived *not* to exist. Clearly, such a being is greater than one we can think of as possibly not existing. If a being can be conceived not to exist, however, he is definitely *not* a being than which no greater can be conceived. "But this is an irreconcilable contradiction. There is, then, so truly a being than which nothing greater can be conceived to exist, that it cannot even be conceived not to exist; and this being you are, O Lord, our God."[8]

Dawkins treats Anselm's argument not to analysis but to ridicule:

> Let me translate this infantile argument into the appropriate language, which is the language of the playground.
>
> "Bet I can prove that God exists."
>
> "Bet you can't."
>
> "Right then, imagine the most perfect, perfect, *perfect* thing possible."
>
> "Okay, now what?"
>
> "Now, is that perfect, perfect, *perfect* thing real?"
>
> "Does it exist?"
>
> "No, it's only in my mind."
>
> "But if it was real it would be even more perfect, because a really, really perfect thing would have to be better than a silly old imaginary thing. So I've proved that God exists. Nur Nurny Nur Nur. All atheists are fools."[9]

While the consensus follows Dawkins in thinking that Anselm's ontological argument is flawed, and while there have been many refutations over the years, all the way up to the present,[10] that there are *many* refutations is just the problem. The consensus that the argument is flawed has not been accompanied by a consensus that identifies what exactly flaw is. Some might think that an argument that has survived such scrutiny remains alive and well.

There is another point about Anselm's argument that Dawkins fails understand: "An odd aspect," he says," is that it was originally addressed not to humans but to God himself, in the form of a prayer (you'd think that any entity capable of listening to a prayer would need no convincing of his own existence)."[11] Dawkins does not understand that Anselm *already* believes in God, a matter of faith, not reason. Anselm is not trying to convince anyone that God exists. Reason is no substitute for faith; it can only support it, as Anselm makes clear in his preface.

Is the Bible Worth Reading?

Dawkins does not argue only against God; he also argues against the Bible as an ethical source. An analogy with the musical satire of Anna Russell may help us understand Dawkins's view of this central document in the spiritual history of the West. Her *"The Ring of the Nibelungs:* An Analysis" is an astonishing lampoon, an insightful tour of a music drama chock-full of absurdities even Richard Wagner could not take seriously. These include a lover's meeting that takes place in a living room tastefully decorated with a sword plunged into an ash tree. The sum of this accumulated nonsense justifies her immortal signature line: "I'm not making this up, you know!"

But there the parallel ends. Beneath the laughter, Russell and her audience share a judgment: Wagner's *Ring* is perhaps the greatest music drama ever composed, a profound enactment of the eternal conflict between power and love. While their laughter is tinged with respect, even awe, Dawkins's view of the Old and New Testaments is derisive, a derision hard to understand. For millennia, despite persistent persecution, the Jews managed to maintain their identity as a people while once proud empires disappeared without a trace, not excluding the thousand-year Reich. Still, Dawkins is not impressed. Moreover, he attributes to gullibility the fact that a single Jew, born in poverty in an obscure corner of a great empire and executed as a criminal after a short career, permanently changed the world.

According to Dawkins our morality, our sense of right and wrong, comes not from the Bible but from our selfish genes:

> We now have four good Darwinian reasons for individuals to be altruistic, generous and "moral" toward each other. First, there is the special case of genetic kinship. Second, there is reciprocation: the repayment of favors given, and the giving of favors in "anticipation" of payback. Following on from this there is, third, the Darwinian benefit of acquiring a reputation for generosity and kindness. And fourth, if [animals can assert their dominance through acts of generosity and risk-taking on behalf of others], there is a particular additional benefit of conspicuous generosity as a way of buying unfakeably authentic advertising.[12]

Indeed, altruism may be analogous to sexual desire; we may have "a lust to be generous,"[13] a passion that carries over from ancestral village life bound by ties of kinship to the modern world where most of those we meet are strangers. We have to be nice to others; we simply cannot help ourselves.

According to Dawkins, this correct view contrasts with the explanation that we get our ethics from the Bible, a view that on closer scrutiny does not hold water, because "the God of the Old Testament is arguably the most unpleasant character in all fiction; jealous and proud of it; a petty, unjust control-freak."[14] Moreover, the "exemplary" patriarchs, Abraham and Moses, are, as the Bible makes perfectly clear, pretty disreputable.[15] His view of the sacrifice of Isaac is typical. It is "child abuse," by a supposedly loving father and a loving God, an example of "the Nuremberg defense with Abraham cast as the Nazi just following orders. And it does no good to say it's just a story, a fairy tale with a moral. In the first place, many people believe it isn't just a story. In the second place, what could the moral be? Surely nothing praiseworthy."[16] But to judge in this way is seriously to misjudge a tale that over the centuries has troubled and inspired the fathers of the Church, the anonymous authors of the medieval mystery cycles, the philosopher Max Horkheimer, the literary critic Eric Auerbach, the theologian Jack Miles, and the existential philosopher Søren Kierkegaard, a tale that has inspired artists and musicians throughout the ages.

For the Church fathers and for the authors of the medieval mystery plays, the sacrifice of Isaac foreshadowed the Crucifixion.[17] In his polemic against the Manichean Faustus, Augustine makes clear the popularity of the tale[18] and its agreed-upon meaning:

> You say that Christ was not foretold by the prophets of Israel, when, in fact, their Scriptures teem with such predictions, if you would only examine them carefully, instead of treating them with levity. Who in Abraham leaves his country and kindred that he may become rich and prosperous among strangers, but He who, leaving the land and country of the Jews, of whom He was born in the flesh, is now extending His power, as we see, among the Gentiles? Who in Isaac carried the wood for His own sacrifice, but He who carried His own cross? Who is the ram for sacrifice, caught by the horns in a bush, but He who was fastened to the cross as an offering for us?[19]

This meaning of the sacrifice of Isaac is central to the Abraham and Isaac plays of the medieval mystery cycles, cycles in which as in Augustine the Old and New Testaments interpenetrate and fuse:

> This deed you've seen done here in this place,
> In example of Jesus done it was,
> That for to win mankind grace
> Was sacrificed on the rood.

By Abraham I may understand
The Father of heaven that can find
With his Son's blood to break that bond
What the devil had brought us to.
By Isaac understand I may
Jesus that was obedient aye,
His Father's will to work always
And death for to confound.[20]

To Max Horkheimer, actions like those of Abraham, obedient to God without question, are absolutely central to Jewish life. They are a form of immortality without an afterlife, realized in the continuity of tradition, the succeeding generations of Jews whose daily activities are a form of worship:

> The idea of continued existence signifies in the first instance, not the after-life, but the identification of the nation so crassly distorted by modern nationalism, which has its prehistory in the Bible. By conducting his life in accordance with the Torah, by spending days, months, years, in obedience to the Law, the individual becomes so much one with others despite personal differences that after his death he continues to exist through those who survive him, in their observance of tradition, and love for the family and the tribe, in the expectation that at some time things may still become better in the world. . . . Not unlike the figure of Jesus in Christianity, Judaism *as a whole* bore witness to redemption.[21]

The literary criticism of Erich Auerbach illustrates some of the ways in which so brief a tale as that of Abraham and Isaac can make so lasting an impression on Western culture. Auerbach analyzes a narrative bereft of detail, deliberately so, because its author intends to focus entirely on the relationship between God and Abraham, a man whose very life an act of worship, a Jew on the model proposed by Max Horkheimer. Of the Old Testament story we may ask: Where is God? Where is Abraham? We are not told. In fact, they meet not in real but in ethical space. Because this meeting is the occasion for God's shocking command, the story of the sacrifice of Isaac demands interpretation. It seems unethical for God to issue such a command; equally, it seems unethical for Abraham to accede. But Abraham obeys without question, indeed with alacrity: he starts out *early* the next morning on a journey he cannot explain his wife, an event whose purpose he cannot share with his wife on his return.

We may be inclined to treat this encounter between the human and divine as an exemplary fiction. But, as Auerbach points out, it is presented as historical fact; indeed, it must be so intended:

> The Bible's claim to truth . . . is tyrannical—it excludes all other claims. The world of the Scripture stories is not satisfied with claiming to be a historically true reality—it insists that it is the only real world, is destined for autocracy. All other scenes, issues, and ordinances have no right to appear independently of it, and it is promised that all of them, the history of all mankind, will be given their due place within that frame, will be subordinated to it.[22]

It is just this tyranny that literary critic Jack Miles denies. He calls the sacrifice of Isaac a fable and searches the text for signs of ambiguity to convince us that Abraham may never have intended to carry through with the sacrifice of his son. Doesn't Abraham say to his servants that both he and Isaac will return after the sacrifice? Doesn't he say that "God will provide Himself the lamb for the burnt offering"? Saying this, is he speaking to Isaac or to God? And if to God, is this a plea? A challenge? For the God of Jack Miles can be challenged: the encounter between him and Abraham is not an example of absolute obedience but an exercise in mutual "self-discovery."[23] Both discover that obedience to God must correspond to the moral law.

For Søren Kierkegaard the sacrifice of Isaac is neither the only true account of Auerbach nor the fable of Miles; it is instead the occasion for a book-length meditation on the nature of faith, a book called *Fear and Trembling*, the two reactions that necessarily accompany such a meditation. The title, drawn from the Epistles, emphatically makes this point: "Work out your own salvation with fear and trembling" (Phil. 2:12). Because for Kierkegaard life without God, without the possibility of salvation, is meaningless:

> If there were no eternal consciousness in a man, if at the foundation of all there lay only a wildly seething power which writhing with obscure passions produced everything that is great and everything that is insignificant, if a bottomless void never satiated lay hidden beneath all—what then would life be but despair? If such were the case, if there were no sacred bond which united mankind, if one generation arose after another like the leafage in the forest, if the one generation replaced the other like the song of birds in the forest, if the human race passed through the world as the ship goes through the sea, like the wind through the desert, a thoughtless and fruitless

activity, if an eternal oblivion were always lurking hungrily for its prey and there was no power strong enough to wrest it from its maw—how empty then and comfortless life would be![24]

For Kierkegaard, the only escape from this despair is a faith so absolute that it denies the bonds of friendship and family. Anyone in search of such a faith must take seriously Jesus's admonition that "if any man come to me, and hate not his father, and mother, and wife, and children, and brethren, yea, and his own life also, he cannot be my disciple" (Luke 14:26). This is because for Kierkegaard we cannot arrive at faith unless we experience the "infinite resignation" that is the subject of Jesus's admonition, the indication that one's duty toward God is absolute. In his willingness to sacrifice the person dearest to him, Abraham is exemplary of this faith. His act represents "the teleological suspension of the ethical" in the interest of this absolute duty: "The ethical expression for what Abraham did is, that he would murder Isaac; the religious expression is, that he would sacrifice Isaac; but precisely in this contradiction consists this dread which can well make a man sleepless, and yet Abraham is not what he is without this dread."[25]

Both Kierkegaard and Miles comment on the same biblical passage: "God will provide Himself the lamb for the burnt offering." To Kierkegaard, though not to Miles, the sentence is ironic; it equivocates, saying nothing whatever about Abraham's mission. Saying this, Abraham is able to keep his secret, a condition necessary for maintaining a faith free of ethical concerns. Surely Isaac and Sarah would challenge him on this ground. But Abraham's saying is nevertheless infused with hope. For it is possible after all that Isaac will not be the lamb, "that God could do something entirely different."[26]

We turn now from what the story means to what it has inspired. First, we focus on the Chester *Abraham and Isaac*. In this play, we see not the stoic figures of the biblical account, but a father in anguish:

ISAAC. Ay mercy, father, why tarry ye so?
Smite of my head and let me go.
I pray you rid me of my woe,
For now I take my leave.

ABRAHAM. My son, my heart will break in three
To hear thee speak such words to me
Jesus, on me thou have pity,
That I have most of mind.[27]

FIGURE 13.1 The sacrifice of Abraham and Isaac. Stained glass panel in Canterbury Cathedral.

SOURCE: Sonia Halliday Photo Library.

The sacrifice of Abraham and Isaac inspired not only medieval drama but the fine arts of the Middle Ages. In the Canterbury stained glass panel in figure 13.1, our habit of reading from left to right leads us from the ram to the angel. But to Abraham, about to strike the fatal blow, the two are out of sight. At last, we arrive at Isaac, trussed up and positioned against a bundle of firewood in the shape of a cross, clearly a type of Christ, the Old and New Testament fused in a single image.

The contrast between the Canterbury depiction and one from the modern era, by Rembrandt, is dramatic. In figure 13.2, the Dutchman depicts the instant of divine intervention, the knife in midair, falling from the interrupted hand. To emphasize a key structural feature of the painting, an inverted triangle whose apex is Isaac's throat, light from the heavens pours toward that point. With Rembrandt, we have moved from the static to the dynamic, from the typological to the human.

Two 20th-century composers, Benjamin Britten and Igor Stravinsky, exploit the expressive potential of Isaac's sacrifice. Britten's *War Requiem*, written to celebrate the dedication of a new cathedral at Coventry, the replacement for one the Nazis destroyed, is a deeply moving pacifist

FIGURE 13.2 Rembrandt, *The Sacrifice of Isaac*.

hymn. Its third movement, the Offertorium, sets a poem by Wilfred Owen that reconfigures the Abraham story with startling intent:

Then Abram bound the youth with belts and strops,
And builded parapets and trenches there,

And stretched forth the knife to slay his son.
When lo! an angel called him out of heaven,
Saying, Lay not thy hand upon the lad,
Neither do anything to him. Behold,
A ram, caught in a thicket by its horns;
Offer the Ram of Pride instead of him.

But the old man would not so, but slew his son,
And half the seed of Europe, one by one.

Although Britten borrows the music accompanying this poem from his earlier *Canticle II: Abraham and Isaac*, he gives it a savage twist. The last line of the poem is repeated by the baritone and tenor singing against the choir. The music of the soloists and the choir does not blend, however, an effect deliberately jarring. While the baritone and tenor sing of the slaughter in the trenches, the choir sings, "Make them, O Lord, pass from death into life."

In his "Abraham and Isaac," Stravinsky sets the Hebrew text, syllable by syllable. The score, dedicated "to the people of Israel," rests in the Israel Museum, which overlooks "the beautiful Byzantine Russian Orthodox Monastery and the eternal hills of the land of Abraham and Isaac."[28] The work is written for a baritone soloist and a chamber orchestra. Lasting just under twelve minutes, it is uncompromisingly difficult, its riches revealed only on repeated hearings. As an example of the significance with which music can infuse the story, there is the point at which the heavens intervene, a juncture marked by a flute and a tuba playing together, a combination symbolizing the compass of God's power.

In his attack on biblical ethics, Dawkins attacks not only the Old but the New Testament. To be sure, in his view the New Testament is an improvement: "Jesus . . . was surely one of the great ethical innovators of history. The Sermon on the Mount is way ahead of its time."[29] Still, Jesus proves Dawkins's point. He was an innovator: he departed from biblical ethics. After all, didn't he contravene these when he said in Mark that the Sabbath could be ignored, that the Sabbath was made for man, not man for the Sabbath? But Jesus was most assuredly not an innovator in Dawkins's sense: he thought of himself as fulfilling the Law, not abandoning it. On adultery, the Law of Moses is clear: the sinner is to be stoned to death. But when a woman taken in adultery is brought before him, what does Jesus do? He dismisses her accusers, saying, "He that is without sin among you, let him cast the first stone at her" (John 8:7).Stung by a level of moral insight far above their own, they leave the scene. Jesus then turns to the woman: "Where are those thine accusers? Hath no man condemned thee?

She said, No man, Lord. And Jesus said unto her, Neither do I condemn thee: Go and sin no more" (John 8:10–11).

Several things need to be said about this remarkable exchange. The first is that Jesus, the accusing scribes and the Pharisees, and the woman taken in adultery exist in the same moral universe. Jesus does not contravene the Law; he interprets it. Adultery remains a sin, but Jesus condemns the sin, not the sinner. Instead, he forgives the sinner, recognizing that we all need to be forgiven. What is transformed by his example is not only our sense of justice but our sense of what it is to be an upright human being.

None of this is to deny Dawkins's view that kinship and reciprocity play a role in forming human relationships. It is only to deny that these factors are sufficient to account for the ethical insights of the Old and the New Testament and their interpreters. These factors are especially inadequate to explain why our sense of right and wrong should on occasion overrule the rules, just as Jesus does. Dawkins seems to agree that his selfish genes cannot account for all our apparently good behavior:

> When I say something like "A child should lose no opportunity for cheating . . . lying, cheating, deceiving, exploiting . . ." I am using the word "should" in a special way. I am not advocating this kind of behavior as moral or desirable. I am simply saying that natural selection will tend to favor children who do act in this way, and that therefore when we look at wild populations we may expect to see cheating and selfishness within families. The phrase "the child should cheat" means that genes that tend to make children cheat have an advantage in the gene pool. If there is a human moral to be drawn, it is that we must *teach* our children altruism, for we cannot expect it to be part of their biological nature.[30]

But according to Dawkins our biological nature formed by natural selection is all that we have. What then can be the source of the altruism we are to teach to our children?

Must Science and Religion Be at Odds?

There is a consensus that the rise of science in the Middle Ages is unthinkable in the absence of Islam and of Christianity. Concerning Arabic science, A. I. Sabra feels that "we have to do with a single unitary tradition," one that was "the accomplishment of individuals who experienced the intersection, at a certain place and time, of three major movements at work—namely those of Hellenism, Arabism, and Islamism."[31] Concerning Western science, Jürgen Habermas opines that "the modern forms of consciousness encompassing abstract right, modern science, and autonomous art (with the secularization and independence of the panel painting at its center) could never have developed apart from the organizational forms of Hellenized Christianity and the Roman Catholic Church, without the universities, monasteries, and cathedrals."[32]

In the Middle Ages, science progressed within the contexts of Islam and Catholic Christianity. The works of Greek science had already been translated and fruitfully employed in the Arab lands when, beginning in the 11th century, a wave of translations into Latin, the lingua franca of learning, introduced Aristotle and Ptolemy to the West. But the West did not simply absorb Greek and Arabic science. Departing from Aristotle, Jean Buridan developed the concept of impetus, the theory that a thrower imparts a force to the projectile, one that does not diminish unless a contrary force is applied. Nicole Oresme produced the best-known proof of the mean-speed theorem, possibly the most outstanding achievement of medieval science. According to this theorem, a uniformly accelerated body travels the same distance as a body with uniform speed whose speed is half the final velocity of the accelerated body.[33]

To say that the rise of science in the West and in the Arabic lands is unthinkable without Christianity and Islam is not to say that the marriage of religion and science was always happy. What marriage is? In 1277, the Church came down hard on the doctrine of double truth, the position that because the truths of religion and science move in separate paths, they cannot in principle contradict each other. Thus, when science deviated from divine truth, no heresy can have been committed. In 1277, the pope undermined this principle, condemning on theological grounds 219 propositions scientists were forbidden to hold.[34]

Another uncomfortable time for science followed from Copernicus's insistence that his system was not just a mathematical convenience. In the words of Thomas Kuhn:

> More than a picture of the universe and more than a few lines of Scripture were at stake. The drama of Christian life and the morality that had been made dependent on it would not readily adapt to a universe in which the earth was just one of a number of planets.[35]

Today, most religious creeds accept that science reigns supreme in the natural world; the past antagonism with science is forgotten. The sole dissent from this consensus arises from any creed that insists on a literal interpretation of Genesis in particular and of Scripture in general. Evolutionary theory is the usual target of such creeds. To scientists like Dawkins, to discard evolutionary theory on the basis of this interpretation of Scripture is to participate in the triumph of dogma over truth.

While I must perforce agree, in the case of evolution, scientists themselves may be victims of dogma, the dogma that all legitimate explanations must of necessity be materialistic and naturalistic. While the views of the Young Earth Creationists and Creation Scientists play fast and loose with evidence that clearly contradicts their claims, it is untrue that standard evolutionary theory, the evolutionary theory of Carson, Dawkins, Gould, Pinker, and Wilson, is invulnerable to plausible counterargument, to the charge that standard theory is more dogma than truth in that its underlying naturalism and materialism does not make *scientific* sense.

Alvin Plantinga, a theist, questions whether natural selection can account for our beliefs. He concedes that these are adaptive, that their truth is important to our survival. But he objects to using this obvious fact to argue in favor of a naturalistic and materialistic theory of evolution. To do this, he feels, is to beg the question. If natural selection were sufficient to explain who we are, we would be in the same position as cockroaches and canaries, turning perceptions directly into behavior; we would not need beliefs any more than they do. For natural selection, beliefs are superfluous; they have nothing to do with reproductive fitness.[36] Thomas Nagel, an atheist, also questions whether any materialistic theory is sufficient to account for the fact that we are conscious, that we think, and that our behavior is guided by values. It is unarguable that these characteristics are not material. But if this is so, there must, he feels, be "a cosmic predisposition to the formation of life, consciousness, and the value that is inseparable from them."[37] While Nagel dissents from Plantinga's position that a naturalistic theory cannot account for the fact that we have beliefs, he concurs with philosopher Sharon Street that

> if the responses and faculties that generate our value judgments are in significant part the result of natural selection, there is no reason to expect that they would lead us to be able to detect any mind-independent moral or evaluative truth, if there is such a thing. This is because the ability to detect such truth, unlike the ability to detect mind-independent truth about the physical world, would make no contribution to reproductive fitness.[38]

It is especially impressive when theists and atheists agree. But the reader does not have to find these arguments compelling to concur with me that they cannot be dismissed with sarcasm and a wave of the hand. None of this, of course, argues against the compatibility of science and religion. It does suggest, however, that scientists may be overly hasty in dismissing religion—and philosophy—as irrelevant to the task of understanding the natural world.

What Have They Got Against God?

The courts have spoken. After their decisions in *McLean v. Arkansas* and *Edwards v. Aguillard*, science, not religion, is the official arbiter of knowledge of the natural world. Creation science is, definitively, not science,[39] and the teaching of Genesis is barred from the public schools. Nevertheless, any assertion of the triumph of scientific over biblical authority is seriously misleading. Despite the acceptance of evolution by the Catholic Church and most Protestant denominations, "forty-six percent of Americans believe in the creationist view that God created humans in their present form at one time within the last 10,000 years." Indeed, the prevalence of this creationist view of the origin of humans is essentially unchanged from 30 years ago, when Gallup first asked the question.[40] If we focus on those who attend church weekly, this figure jumps to 67%. Furthermore, 31% of Americans think that even scientists have serious doubts about evolution.[41] While three-quarters of Americans would not be upset if creationism were taught in the public schools, three in ten would be upset if only evolution were taught.[42] Our scientist-popularizers might well feel angry at a God responsible for these serious and persistent misapprehensions, misapprehensions most creeds do not endorse but that a majority of their believers apparently do.

For *McLean* solved nothing, nor was the Supreme Court's decision in *Edwards* a crushing blow: court cases followed in 1990, 1994, 1997, 2000, 2005, and 2006.[43] Why were *McLean* and *Edwards* unsuccessful? Why was even the Supreme Court unsuccessful in bringing an end to lawsuits in the lower courts? This is not *Brown v. Board of Education*, a decision successfully backed with the force of the federal government. What high school science teachers teach depends not on the courts but on their personal beliefs, the beliefs of their communities, and a level of preparation that is, in many cases, far from adequate.[44] Sixteen percent of high school biology teachers believe that "God created human beings pretty much in their present form at one time within the last 10,000 years or so"; only 28% believe that "human beings developed over millions of years . . . but God had no part in the process."[45] These facts are far from conducive to making evolutionary theory the organizing principle to high school courses in biology.

While it is true that America is nearly the worst offender when it comes to a disbelief in evolution and a belief in Genesis as science, other countries are not free from this illusion, as figure 13.3 demonstrates.

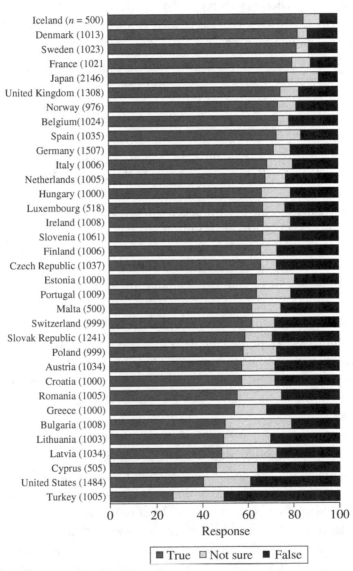

FIGURE 13.3 Public Acceptance of evolution.

SOURCE: Jon D. Miller, Eugene C. Scott, and Shinji Okamoto, "Public Acceptance of Evolution," *Science* 313 (August 11, 2006): 765–66.

Indeed, we might be right in thinking that matters have begun to move in the direction of increasing ignorance. In 2007, by a vote of 48–25 with three abstentions, the Parliamentary Assembly of the Council of Europe passed an even-handed resolution against creationism:

> The aim of this resolution is not to question or to fight a belief—the right to freedom of belief does not permit that. The aim is to warn against certain tendencies to pass off a belief as science. It is necessary to separate belief from science. It is not a matter of antagonism. Science and belief must be able to coexist. It is not a matter of opposing belief and science, but it is necessary to prevent belief from opposing science.

The endorsement of the Parliamentary Assembly, however, was hardly sweeping. Delegates from Armenia, Belgium, the Czech Republic, Estonia, Germany, Iceland, Italy, Lithuania, Moldova, the Netherlands, Norway, Poland, the Russian Federation, Serbia, and the Slovak Republic were opposed.[46]

When we journey beyond Europe, the news is equally dismal. The Turks have gone farther than the chart indicates: "It is not only high school education that is affected," biology professors Zehra Sayers and Zuhal Özcan explain. "There are no universities in Turkey offering undergraduate or graduate degrees in evolutionary biology or in related fields, and even courses in the area are hard to come by." They add that "there is plenty more to ensure that the picture is truly bleak: the circulation of glossy anti-evolution books in the country and abroad . . . , the organization of anti-evolutionary symposia . . . , and the banning of student-organized discussion panels on evolution in some universities." The South Koreans, not represented in the chart, have even excised evolution from their science textbooks.[47]

Figure 13.4 shows that Britain, America's complement as a source of books of popular science, is far less wedded than the United States to creationism and intelligent design. This does not mean, however, that Britain is free from the effects of science miseducation. Among the upper middle class and the middle class, only 70% have heard of Charles Darwin and know a lot or a fair amount about his theory of evolution; among the working class and the casually employed, pensioners, and welfare recipients a mere 25% are in that category. When it comes to science British schools are failing the middle class; schools for those on the lower rungs of society are achieving little or nothing.

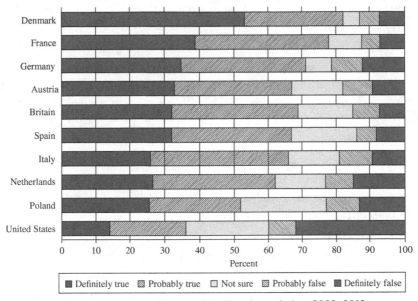

FIGURE 13.4 Percentage who believe or disbelieve in evolution, 2002–2003.

SOURCE: Jon D. Miller, Eugene C. Scott, and Shinji Okamoto, "Public Acceptance of Evolution," *Science* 313 (August 11, 2006): online supplemental materials.

A deep commitment to biblical literalism is hardly the problem: 15% of British Christians have never read the Bible; only 35% identified Matthew as the first book of the New Testament; and only 44% believe that Jesus is the son of God, the Savior of humankind. Despite this, a mere 14% feel that the Genesis account of origins should not be taught in state-funded schools.[48] This preference chimes in well with the views of the teachers themselves. When asked if creationism should be mentioned, but only to demonstrate that no scientific evidence supports it, over half the teachers disagree.[49]

What accounts for the enduring power of Genesis? It is a narrative designed to move us deeply by the awe it inspires at the vastness of God's creation and the frisson it generates at our realization of a power that could will the universe into existence. It is a passage Longinus cites with approval:

In the beginning God created the heaven and the earth. And the earth was without form, and void; and darkness was upon the face of the deep. And the Spirit of God moved upon the face of the waters. And God said, Let there be light: and there was light. And God saw the light, that it was good: and God divided the light from the darkness. And God called the light Day, and the darkness he called Night. And the evening and the morning were the first day. (Gen. 1:1–5)

An advertisement for a data storage system presents us with an altogether different vision of our origin, a view of the sublime shared by many scientists and explicitly or implicitly by all the science writers with whom we deal:

> The Higgs Boson, better known as "the God particle," is what holds the basic building blocks of our universe together. For years, scientists have been trying to isolate it with a Large Hadron Collider (LHC) that creates 6,000,000 collisions among 3,000,000,000,000,000 [3 quadrillion!] protons every second. In the same time, over 50,000,000,000,000 [50 trillion!] bytes of data are being analyzed to prove the particle's existence.[50]

Popular science writers are motivated not only by science miseducation and public ignorance but by their own growing atheism, their increasing sense of the irrelevance of God. According to the National Association of Science, "whether or not God exists is a question about which science is neutral." Historian of science Ronald Numbers points out that the Association reached this conclusion only by "ignoring centuries of history."[51] Leading scientists are not neutral on this issue: in 1998 only 7% believed in a personal God, down from 27.7% in 1914.[52] According to another survey, one in four elite scientists and social scientists described themselves as believers in a spirituality that "meshes beautifully with their identities as scientists because they also see spirituality as an individual journey, as a quest for meanings that can never be final, just as is the case for scientific explanations of reality. For some their sense of spirituality flows very deeply from the work that they do as scientists."[53] One scientist says:

> I don't belong to any religion now. I always assume that people who have spirituality believe in God and they think of it that way. Personally I believe in nature, and I get my spirituality, if you want to say that, from being in nature, but I don't really believe there's a God, so I don't consider it's necessary for what I do or how I behave.[54]

In his Nobel lecture on the discovery of buckminsterfullerene, a new form of carbon, Richard Smalley reiterates this creed: "This discovery was one of the most spiritual experiences that any of us in the original team of five have ever experienced. The main message of my talk today is that this spiritual experience, this discovery of what Nature has in store for us with carbon, is still ongoing."[55]

Theirs is a universe in which God plays no part, one in which science plays the part of God. This substitution is necessary because the laws of physics and evolutionary theory seem to them wholly incompatible with a belief in God the designer, a consequence first noted by the scientist and Christian Robert Boyle:

> The excessive veneration men have for nature, as it has made some philosophers [scientists] . . . deny God, so it is to be feared that it makes many forget him. . . . The erroneous idea of nature would, too often, be found to have a strong tendency to shake, if not subvert the very foundations of all religion; misleading those, that are inclined to be its enemies, from overlooking the necessity of a God, to questioning, if not to the denial of his existence.[56]

It is the stubborn persistence of a personal God in people's lives—a God who trumps the findings of science concerning origins—that accounts for the antagonism to religion that characterizes the work of E. O. Wilson, Richard Dawkins, Stephen Hawking, Steven Pinker, and Steven Weinberg and the indifference to religion of Rachel Carson, Lisa Randall, Brian Greene, Stephen Jay Gould, and Richard Feynman. It is Feynman who gives Boyle's fear its most eloquent expression:

> We have been led to imagine all sorts of things infinitely more marvelous than the imaginings of poets and dreamers of the past. It shows that the imagination of nature is far, far greater than the imagination of man. For instance, how much more remarkable it is for us all to be stuck—half of us upside down—by a mysterious attraction to a spinning ball that has been swinging in space for billions of years than to be carried on the back of an elephant supported on a tortoise swimming in a bottomless sea.[57]

Conclusion

God's existence can neither be proved nor disproved. There is no evidence for or against this hypothesis because it is not a hypothesis; it is an article of faith. Like theists, atheists depend on faith for their conviction that God is a delusion. But faith is something that, one would think, scientists behaving as scientists would strenuously avoid. It would be distinctly odd for Newton, a believer, to say he had faith that space was absolute or for Einstein, a nonbeliever, to say that he had faith in the curvature of space-time. Where God is concerned, scientists would, one would think, act like scientists; they would suspend judgment; they would be agnostics. They would also realize that as scientists they are not equipped to argue against God. No scientists can claim that their arguments against God or the authority of the Bible would pass muster by the standards they and their fellows routinely uphold in their own work.

In making arguments that would pass muster by those high standards, scientists would review the evidence first; then they would check and double-check whether that evidence supported a claim they favored, one on which they could safely stake their reputations. They would behave like searchers for the truth; if they did so, they would have to conclude that science can say nothing about the meaning of life or of the good life. They would admit that their kind of smart isn't the only kind: they would also admit that to sneer at the likes of Plato, Anselm, Descartes, Horkheimer, Auerbach, Kierkegaard, Miles, Habermas, Sabra, Plantinga, or Nagel— to sneer at what one does not trouble to understand—puts them exactly in the position of those who prefer biblical accounts of creation to the scientific truth. It is Dawkins himself who warns us against such cross-disciplinary pontificators: "Publishers should correct the misapprehension that a scholar's distinction in one field implies authority in another. And as long as that misapprehension exists, distinguished scholars should resist the temptation to abuse it."[58]

NOTES

Chapter 1

1. Albert Einstein, *Ideas and Opinions*, (New York: Crown, 1954), 11.

2. "Nous ravit tous d'admiration et d'étonnement et nous surprit de telle sorte, que, pour notre satisfaction propre, nous voulûmes la répéter." From Blaise Pascal, *Oeuvres Complètes*, vol. 2 (Paris: Librairie de L. Hachette, 1858), 314. Mine is a free translation of a sentence from a letter sent to Blaise Pascal by his brother-in-law, Florin Périer, reporting an experiment that confirmed that air had weight. Périer had tested his apparatus at various heights on the extinct French volcano Puy de Dôm.

3. Elizabeth Leane, *Reading Popular Physics: Disciplinary Skirmishes and Textual Strategies* (Aldershot, UK: Ashgate, 2007).

4. Magdalena Ball, "*A Brief History of Time* by Stephen Hawking," *Blogcritics*, February 1, 2009, http://blogcritics.org/book-review-a-brief-history-of.

5. Jon Turney, "Commentary for *Public Understanding of Science* Special Issue," *Public Understanding of Science* 22, no. 5 (2013): 571.

6. Jon Turney, "The Abstract Sublime: Life as Information Waiting to be Rewritten," *Science as Culture* 13, no. 1 (2004): 91.

7. Marjorie Hope Nicolson, *Mountain Gloom and Glory: The Development of the Aesthetics of the Infinite* (Seattle: University of Washington Press, 1959).

8. Quoted in James Secord, *Victorian Sensation: The Extraordinary Publication, Reception, and Secret Authorship of* Vestiges of the Natural History of Creation (Chicago: University of Chicago Press, 2003), 418.

9. Ibid., 3.

10. Bernard Lightman, *Victorian Popularizers of Science: Designing Nature for New Audiences* (Chicago: University of Chicago Press, 2007), 412–13.

11. Quoted in Ian Hesketh, "The Recurrence of the Evolutionary Epic," *Journal of the Philosophy of History* 9 (2015): 202.

12. Edward O. Wilson, *On Human Nature* (Cambridge, MA: Harvard University Press, 2004), 192.

13. M. Eger, "Hermeneutics and the New Epic of Science," in *The Literature of Science: Perspectives on Popular Scientific Writing*, ed. Murdo William McRae (Athens, GA: University of Georgia Press, 1993), 191, 198.

14. Jeff Hester, quoted in Elizabeth A. Kessler, *Picturing the Cosmos: Hubble Space Telescope Images and the Astronomical Sublime* (Minneapolis: University of Minnesota Press, 2012), 225.

15. David E. Nye, *Electrifying America: Social Meanings of a New Technology* (Cambridge, MA: MIT Press, 1991); David E. Nye, *American Technological Sublime* (Cambridge, MA: MIT Press, 1994); Leo Marx, *The Machine in the Garden* (New York: Oxford University Press, 2000), 197.

16. Longinus, *On the Sublime*, translated by W. H. Fyfe (Cambridge, MA: Harvard University Press, 1995), 163.

17. Ibid., 163.

18. Ibid., 225.

19. Ibid., 277.

20. Bernard Weinberg, "Translations and Commentaries on Longinus, 'On the Sublime,' to 1600: A Bibliography," *Modern Philology* 47, no. 3 (2013): 145–51.

21. Andrew Ashfield and Peter de Bolla, *The Sublime: A Reader in British Eighteenth-Century Aesthetic Theory* (Cambridge: Cambridge University Press, 1996), 37.

22. Ibid., 39.

23. Homer, *Homer's Illiad*, trans. Alexander Pope, ed. W. C. Armstrong (New York: World, 1877), 433.

24. *Aristotle: Poetics; Longinus: On the Sublime; Demetrius: On Style* (Cambridge, MA: Harvard University Press, 1996), 147.

25. Ibid., 155–56.

26. Ibid., 47.

27. Ibid., 94.

28. Longinus, *On the Sublime*, 153.

29. Brian William Cowan, "Mr. Spectator and the Coffee House Public Sphere," *Eighteenth Century Studies* 37, no. 3 (2004): 345–66.

30. E. A. Wrigley, "A Simple Model of London's Importance in Changing English Society and Economy, 1650–1730," *Past and Present* 37 (1967): 44–70.

31. *The Spectator*, with introduction and notes by George A. Aitken, with forty original portraits and eight vignettes (London: Longmans, Green & Co., 1898)), 6: 109.

32. Ibid., 7: 75–76.

33. Longinus, *On the Sublime*, 226–27.

34. Edmund Burke, *A Philosophical Inquiry in the Origin of Our Ideas of the Beautiful and the Sublime* (London: George Bell & Sons, 1889), 10.

35. Ibid., 40.

36. Ibid., 12–13.

37. Ibid., 10–12.

38. Ibid., 37.

39. Longinus, *On the Sublime*, 145.

40. Homer, *Homer's Illiad*, 184.

41. Henry Home of Kames, *Elements of Criticism*, ed. James R. Boyd (New York: A. S. Barnes, 1869), 145.

42. *The Spectator*, 6:76–77; Anthony Earl of Shaftesbury, *Characteristics of Men, Manners, Opinions, Times, etc.*, ed. John M. Robertson (London: Grant Richards, 1900), 112.

43. Thomas Reid, *Inquiry into the Human Mind*, vol. 1 in *The Works of Thomas Reid* (Charlestown, MA: Samuel Etheridge, 1813), 400–401.

44. Immanuel Kant, *Critique of Judgment*, trans. Werner S. Pluhar (Indianapolis: Hackett, 1987), 26.

45. Ibid., 28.

46. Ibid.,23.

47. Ibid., 28.

48. Ibid., 29.

49. Michel de Montaigne, *Complete Works of Montaigne*, trans. Donald M. Frame (Stanford, CA: Stanford University Press, 1967), 1110.

50. Clarence Dewitt Thorpe, "Two Augustans Cross the Alps: Dennis and Addison on Mountain Scenery," *Studies in Philology* 32, no. 3 (1935): 465.*f*

51. Joseph Addison, *Remarks on Several Parts of Italy*, 2nd ed. (London: J.Tonson, 1718), 260–61.

52. Horace-Bénédict de Saussure, *Voyages dans les Alpes*, (Neuchâtel, Switzerland: Fauche-Borel, 1796), 18–19.

53. Joseph Spence, *Letters from the Grand Tour*, ed. Slava Klima (Montreal: McGill—Queens University Press,1975), 111–12.

54. Helen Maria Williams, *A Tour of Switzerland*, vol. 1 (London: G. G. and J. Robinson, 1798), 58.

55. Ibid., 60.

56. Thomas Gray, *The Letters of Thomas Gray*, ed. Duncan C. Tovey, vol. 1 (London: George Bell & Sons, 1909), 44.

57. Ibid., 44

58. Helen Maria Williams, *Tour in Switzerland; or, A view of the present state of the governments and manners of those cantons: with comparative sketches of the present state of Paris* (London. G. G. and J. Robinson, 1798), 1:178.

59. George Keate, *The Alps: A Poem* (London: R. and J. Dodsley, 1758); Williams, *Tour in Switzerland*, 2:16–19.

60. Albrecht von Haller, *Ode sur les Alpes* (Berne: Brounner & Haller, 1773), 46.

61. *The Complete Poems of Thomas Gray: English, Latin and Greek*, ed. Herbert W. Starr and J. R. Hendrickson (Oxford: Oxford University Press, 1966), 151–52.

62. Samuel Taylor Coleridge, *The Collected Works of Samuel Taylor Coleridge, Poetical Works*, vol. 1, ed. J. C. C. Mays (Princeton, NJ: Princeton University Press, 2001), 717–18.

63. Samuel Taylor Coleridge, *Selected Poems*, ed. S. G. Dunn (Oxford: Clarendon, 1918), 112.

64. In regard to his poem, a remark in Coleridge's *Table Talk* is apropos: "Could you ever discover anything sublime, in our sense of the term, in the classic Greek literature? I never could. Sublimity is Hebrew by birth." *The Table Talk and Omniana of Samuel Taylor Coleridge* (London: Oxford University Press, 1917), 191.

65. Marjorie Munsterberg, "J. M. W. Turner's 'Falls at Schaffhausen,' " *Record of the Art Museum, Princeton University* 44, no. 2 (1985): 31.

66. Jeremy Black, *The British Abroad: The Grand Tour in the Eighteenth Century* (Stroud, UK: History Press, 1992), 103.

67. srfab, TripAdvisor review of Air Zermatt, July 2013, http://www.tripadvisor.com/Attraction_Review-g188098-d2281352-Reviews-Air_Zermatt-Zermatt_Valais_Swiss_Alps.html.

68. Don Kaiser, "The Mount Vesuvius Eruption of March 1944," http://www.warwingsart.com/12thAirForce/Vesuvius.html.

69. Pliny the Younger, *Letters of the Younger Pliny*, trans. John Delaware Lewis (London: Trübner., 1879), 192.

70. Ibid., 188.

71. Ibid., 52.

72. Adam Smith, *Essays on Philosophical Subjects*, ed. W. P. D. Wightman and J. C. Bryce (Indianapolis: Liberty Fund, 1982), 39.

73. Ibid., 56.

74. Ibid., 45.

75. Adam Smith, *Theory of Moral Sentiments* (London: George Bell & Sons, 1875), 272.

76. Adam Smith, *Essays*, 105.

77. C. P. Snow, *Two Cultures* (Cambridge: Cambridge University Press, 1998), 72.

Chapter 2

1. Isaiah Berlin, "The Hedgehog and the Fox: An Essay on Tolstoy's View of History," in *The Proper Study of Mankind: An Anthology of Essays*, ed. Henry Hardy and Roger Hausheer (New York: Farrar, Strauss & Giroux, 2000), 493.

2. Lawrence M. Krauss, *Quantum Man: Richard Feynman's Life in Science* (New York: W. W. Norton, 2011), 82, 84.

3. Michelle Feynman, ed., *Perfectly Reasonable Deviations from the Beaten Track: The Letters of Richard Feynman* (New York: Basic Books, 2005), 198–201.

4. Anne Roe, *The Making of a Scientist* (New York: Dodd, Mead, 1953), 58.

5. Joseph C. Hermanowicz, *The Stars Are Not Enough: Scientists, Their Passions and Professions* (Chicago: University of Chicago Press, 1998), 156.

6. Joseph C. Hermanowicz, *Lives in Science: How Institutions Affect Academic Careers* (Chicago: University of Press, 2009), 214.

7. James Gleick, *Genius: Richard Feynman and Modern Physics* (London: Abacus, 1992), 270–71; Krauss, *Quantum Man*, 154–56.

8. A. Zee, introduction to Richard P. Feynman, *QED: The Strange Theory of Light and Matter* (Princeton, NJ: Princeton University Press, 2006), viii.

9. Feynman, *QED*, 3.

10. Richard P. Feynman, *What Do You Care What Other People Think? Further Adventures of a Curious Character* (New York: Bantam Books, 1989), 237.

11. Robert P. Crease and Peter Pesic, "The Feynman Lectures: Fifty Years On," *Physics in Perspective* 16 (2014): 143–45.

12. Feynman, *QED*, 149.

13. Ibid., 24.

14. Ibid., ix–x.

15. Feynman, *Letters*, 98.

16. Feynman, *QED*, 40.

17. Ibid., 45.

18. Ibid., 110.

19. Ibid., 113.

20. Feynman, *Letters*, 408.

21. Feynman, *QED*, 128

22. Dan Styer, "Calculation of the Anomalous Magnetic Moment of the Electron," available online at http://www.oberlin.edu/physics/dstyer/StrangeQM/Moment.pdf.

23. Edmund Burke, *A Philosophical Inquiry in the Origin of our Ideas of the Beautiful and the Sublime* (London: George Bell & Sons, 1889), 52.

24. Richard P. Feynman, *The Character of Physical Law* (Cambridge, MA: MIT Press, 1965), 26.

25. Ibid., 43.

26. Ibid., 53.

27. Ibid., 54.

28. Ibid., 67.

29. Ibid., 94.

30. Ibid., 33.

31. Ibid., 39.

32. Ibid., 129.

33. Ibid.,116.

34. Ibid., 126.

35. Ibid., 126.

36. Quoted in Gleick, *Genius*, 11.

37. Ibid.

38. Richard P. Feynman, *Surely You're Joking, Mr. Feynman! Adventures of a Curious Character* (New York: Bantam, 1985), 8.

39. Ibid., 157–58.

40. Ibid., 229.

41. R. P. Feynman and M. Gell-Mann, "Theory of Fermi Interaction," *Physical Review* 109 (1958): 193–198.

42. Feynman, *Surely*, 312.

43. Feynman, *What*, 180.

44. Ibid., 182.

45. Ibid., 140.

46. Ibid., 148.

47. Ibid., 131.

48. Christopher Sykes, ed., *No Ordinary Genius: The Illustrated Richard Feynman* (New York: W. W. Norton, 1994), 210–14.

49. Edward R. Tufte, *Visual Explanations: Images and Quantities. Evidence and Narrative* (Cheshire, CT: Graphics, 1997), 50–52.

50. Feynman, *What*, 224.

51. Feynman, *Letters*, 108.

52. Feynman, *What*, 18.

53. Feynman, *Surely*, 276–77.

54. Richard P. Feynman, "Banquet Speech," December 10, 1966, available online at http://www.nobelprize.org/nobel_prizes/physics/laureates/1965/feynman-speech.html.

Chapter 3

1. Steven Weinberg, *Facing Up: Science and Its Cultural Adversaries* (Cambridge, MA: Harvard University Press, 2001), 185–186.

2. Steven Weinberg, "A Model of Leptons," *Physical Review Letters* 19 (1967): 1264–66.

3. Ann Finkbeiner, *The Jasons: The Secret History of Science's Postwar Elite* (New York: Penguin, 2006), 155–56.

4. Steven Weinberg, *The First Three Minutes: A Modern View of the Origin of the Universe* (New York: Basic Books, 1993), 4.

5. Weinberg, *First Three Minutes*, viii.

6. Eugene P. Wigner, "The Unreasonable Effectiveness of Mathematics in the Natural Sciences," *Communications in Pure and Applied Mathematics* 13 (1960): 14.

7. Weinberg, *First Three Minutes*, 13.

8. Ibid., 13.

9. Ibid., 168.

10. Edward Hubble, "A Relation between Distance and Radial Velocity among Extra-Galactic Nebulae," *Proceedings of the National Academy of Sciences* 15 (1929): 168–73.

11. Weinberg, *First Three Minutes*, 21.

12. R. H. Dicke, P. J. E. Peebles, P. G. Roll, and D. T. Wilkinson, "Cosmic Blackbody Radiation," *Astrophysical Journal Letters* 1 (1965): 414–19.

13. A. A. Penzias and R. W. Wilson. "Measurement of Excess Antenna Temperature at 4080 Mc/s," *Astrophysical Journal Letters* 1 (1965): 419–21.

14. "Oral History Transcript—Dr. Robert Dicke," American Institute of Physics, June 18, 1985, available online at http://www.aip.org/history/ohilist/4572.html.

15. James Trefil, "The Accident That Saved the Big Bang," *Astronomy* 33 (2005): 40–45; Alan Lightman, "Moments of Truth," *New Scientist* 188 (2005): 36–41; Jim Peebles, "There at the Birth," *New Scientist* 222 (2014): 36–39.

16. "Oral History Transcript—Ralph Alpher," American Institute of Physics, August 12, 1983, available online at http://www.aip.org/history/ohilist/3014_1.html.

17. Weinberg, *First Three Minutes*, 109.

18. Weinberg, *First Three Minutes* 119.

19. Steven Weinberg, *The Discovery of Subatomic Particles* (New York: Scientific American Library, 1983), 13.

20. J. J. Thomson, "Cathode Rays," *Philosophical Magazine* 44 (1897): 293–316.

21. Ibid.

22. G. F. FitzGerald, "Dissociation of Atoms," *The Electrician* 39 (1897): 103–4, my italics.

23. Steven Weinberg, *Dreams of a Final Theory: The Scientist's Search for the Ultimate Laws of Nature* (New York: Pantheon Books, 1992), 4.

24. Ibid., 5.

25. Ibid., 149–50.

26. Ibid., 23.

27. Ibid., 31–32.

28. Roger Penrose, "A Theory of Everything?" *Nature* 433 (2005): 259.

29. Weinberg, *Dreams*, 211.

30. Thomas S. Kuhn, *The Structure of Scientific Revolutions*, 2nd ed. (Chicago: University of Chicago Press, 1970), 170.

31. Weinberg, *Facing Up*, 149.

32. Ibid., 196.

33. Ibid., 200.

34. Ibid., 55.

35. Ibid., 59.

36. P. W. Anderson, "More Is Different," *Science* 177 (1972): 395.

37. "Oral History Transcript—Dr. Philip Anderson," American Institute of Physics, June 29, 2000, available online at http://www.aip.org/history/ohilist/23362_4.html.

Chapter 4

1. Kate McAlpine, "LHC Rap," YouTube, July 28, 2008, http://www.youtube.com/watch?v=j50ZssEojtM.

2. Jim Baggott, *Higgs: The Invention and Discovery of the God Particle* (Oxford: Oxford University Press, 2012), 170.

3. Lisa Randall, *Warped Passages: Unravelling the Mysteries of the Universe's Hidden Dimensions* (New York: Harper Perennial, 2005), 206.

4. Lisa Randall, *The Higgs Discovery: The Power of Empty Space* (New York: Ecco, 2014), 70.

5. Ibid., 73.

6. David E. Nye, *American Technological Sublime* (Cambridge, MA: MIT Press, 1994), 15.

7. Ibid., 85.

8. Hart Crane, "To Brooklyn Bridge," in *The Complete Poems of Hart Crane*, ed. Waldo Frank (New York: Liveright Publishing, 1946), 4.

9. Nye, *American Technological Sublime*, 32, my emphasis.

10. Ibid., xiii.

11. Randall, *Warped Passages*, 127–28.

12. Ibid., 143, my emphasis.

13. Ibid., 129.

14. Ibid., 218.

15. Baggott, *Higgs*, 91.

16. Quoted in Decca Aitkenhead, "Peter Higgs: I wouldn't be productive enough for today's academic system," *Guardian*, December 6, 2013, available online at https://www.theguardian.com/science/2013/dec/06/peter-higgs-boson-academic-system.

17. Lisa Randall, *Knocking on Heaven's Door: How Physics and Scientific Thinking Illuminate the Universe and the Modern World* (New York: Ecco, 2012), 237.

18. Leo Marx, *The Machine in the Garden: Technology and the Pastoral Ideal in America* (Oxford: Oxford University Press, 1964), 214.

19. Frank Wilczek, *A Beautiful Question: Finding Nature's Deep Design* (New York: Penguin, 2015), 227.

20. Randall, *Higgs Discovery*, 2.

21. Randall, *Knocking on Heaven's Door*, 242; Smith, *Essays*, 105.

22. Randall, *Warped Passages*, 77.

23. Ibid., 78.

24. Randall, *Knocking on Heaven's Door*, 247.

25. Ibid., 243.

26. Randall, *Warped Passages*, 156.

27. Randall, *Higgs Discovery*, 2.

28. Randall, *Knocking on Heaven's Door*, 117.

29. Wilczek, *Beautiful Question*, 298.

30. Randall, *Knocking on Heaven's Door*, 302.

31. Brian Greene, *The Elegant Universe: Superstrings, Hidden Dimensions, and the Quest for Ultimate Theory* (New York: Vintage, 2003), 175.

32. Randall, *Knocking on Heaven's Door*, 302.

33. e.e. cummings, *Poems, 1923–1954* (New York: Harcourt Brace, 1954), 397.

34. Randall, *Warped Passages*, 468.

35. Ibid., 250.

36. Lisa Randall and Raman Sundrum, "Large Mass Hierarchy from a Small Extra Dimension," *Physical Review Letters* 83 (1999): 3370–73; Lisa Randall and Raman Sundrum, "An Alternative to Compactification," *Physical Review Letters* 83 (1999): 4690–93.

37. Randall and Sundrum, "Large Mass Hierarchy," 3370.

38. Randall, *Warped Passages*, 250.

39. Ibid., 243–244.

40. Dennis Overbye, "On Gravity, Oreos, and a Theory of Everything," *New York Times*, November 1, 2005, available online at http://www.nytimes.com/2005/11/01/science/on-gravity-oreos-and-a-theory-of-everything.html.

41. Randall, *Warped Passages*, 391.

42. Lisa Randall, libretto to *Hypermusic Prologue: A Projective Opera in Seven Planes*, music by Hèctor Parra (Kairos, 2010).

43. Lisa Randall, *Dark Matter and the Dinosaurs: The Astounding Interconnectedness of the Universe* (New York: Ecco, 2015), xiii.

44. Ibid., 3.

45. Ibid., 288.

46. Ibid., 10.

47. Ibid., 67.

48. Ibid., 320.

49. Ibid., 327.

50. Ibid., 113.

51. Ibid., 206–9.

52. Steven Weinberg, *Dreams of a Final Theory: The Scientist's Search for the Ultimate Laws of Nature* (New York: Vintage, 1994), 37.

53. Ibid., 220.

54. Lisa Randall and Matthew Reece, "Dark Matter as a Trigger for Periodic Comet Impacts," *Physical Review Letters* 112 (2014): 161301; Ji Ji Fan, Andrey Katz, Lisa Randall, and Matthew Reece, "Dark-Disk Universe," *Physical Review Letters* 110 (2013): 211302; Ji Ji Fan, Andrey Katz, Lisa Randall, and Matthew Reece, "Double-Disk Dark Matter," *Physics of the Dark Universe* 2 (2013): 139–156.

55. Randall and Reece, "Dark Matter as a Trigger," 161301–1.

56. Randall, *Dark Matter*, 342.

Chapter 5

1. Lewis Carroll, *The Annotated Alice*, ed. Martin Gardner (New York: W. W. Norton, 2000), 199.

2. R. Easther, B. R.Greene, W. H. Kinney, et al., "Inflation as a Probe of Short Distance Physics," *Physical Review D* 64, no. 10 (November 15, 2001): 103502. 1–8.

3. Ibid.

4. John Horgan, "Troublemaker Lee Smolin Says Physics—and Its Laws—Must Evolve," *Cross-Check* (*Scientific American* blog), January 4, 2015, http://blogs.scientificamerican.com/cross-check/2015/01/04/troublemaker-lee-smolin-questions-if-physics-laws-are-timeless.

5. Peter Woit, *Not Even Wrong: The Failure of String Theory and the Continuing Challenge to Unify the Laws of Physics* (New York: Basic Books, 2006).

6. Sheldon Glashow, profile, Boston University, http://www.bu.edu/khc/about/sheldon-glashow.

7. George Ellis and Joe Silk, "Scientific Method: Defend the Integrity of Physics," *Nature* 516 (2014): 321–23.

8. P. C. W. Davies and Julian Brown, "Richard Feynman," in *Superstrings: A Theory of Everything?* (Cambridge: Cambridge University Press, 1992), 194.

9. Jim Holt, "Unstrung," *New Yorker*, October 2, 2006, 86.

10. John Ellis, "The Superstring: Theory of Everything, or of nothing?" *Nature* 323 (1986): 595.

11. Freeman Dyson, "Science on a Rampage," *New York Review of Books*, April 5, 2012.

12. Tom Siegfried, "Cosmic Inflation has Its Flaws, but So Do Its Critics," *Context* (*ScienceNews* blog), December 11, 2013, https://www.sciencenews.org/blog/context/cosmic-inflation-has-its-flaws-so-do-its-critics.

13. Natalie Wolchover, "In Fake Universes, Evidence for String Theory," *Quanta Magazine*, February 18, 2015, https://www.quantamagazine.org/20150218-string-theory-only-game-in-town.

14. W. H. Newton-Smith, *The Rationality of Science* (London: Routledge & Kegan Paul, 1981), 224.

15. "On the Conflicting Assessments of String Theory," *Philosophy of Science* 76 (December 2009): 995.

16. Brian Greene, *The Fabric of the Cosmos: Space, Time, and the Texture of Reality* (New York: Vintage, 2005), 162.

17. Ibid., xi.

18. Ibid., xi.

19. Brian Greene, *The Elegant Universe: Superstrings, Hidden Dimensions, and the Quest for the Ultimate Theory* (New York: W. W. Norton. 2003), 16.

20. Ibid., 16.

21. Brian Greene, "Why String Theory Still Offers Hope We Can Unify Physics," *Smithsonian Magazine*, January 2015, http://www.smithsonianmag.com/science-nature/string-theory-about-unravel-180953637.

22. Michael B. Green and John H. Schwarz, "Anomaly Cancellations in Supersymmetric D = 10 Gauge Theory and Superstring Theory," *Physics Letters B*, 149 (1984): 117–22.

23. Greene, "Why String Theory."

24. Greene, *Elegant Universe*, 6.

25. Greene, "Why String Theory."

26. Brian Greene, *The Hidden Reality: Parallel Universes and the Deep Laws of the Cosmos* (New York: Vintage, 2011), 176. Excerpt(s) from *The Hidden Reality: Parallel Universes and the Deep Laws of the Cosmos* by Brian Greene, copyright © 2011 by Brian

Greene. Used by permission of Alfred A. Knopf, an imprint of the Knopf Doubleday Publishing Group, a division of Penguin Random House LLC. All rights reserved.

27. Steven Weinberg, *Dreams of a Final Theory: The Scientist's Search for the Ultimate Laws of Nature* (New York: Vintage, 1994), 213.

28. Greene, *Elegant Universe*, 14.

29. Greene, *Fabric*, 332.

30. Edward Abbott, *Flatland: A Romance of Many Dimensions by a Square* (Boston: Roberts Brothers, 1885), 99.

31. *Ibid*, 138–39.

32. Greene, *Fabric*, 14.

33. Brian Greene, Interview with Krista Tippett, *On Being*, January 30, 2014, http://www.onbeing.org/program/brian-greene-reimagining-the-cosmos/transcript/6122.

34. Lawrence M. Krauss, *A Universe from Nothing: Why There Is Something Rather Than Nothing* (New York: Simon & Schuster, 2012), 2.

35. Steven Weinberg, *The First Three Minutes: A Modern Version of the Origin of the Universe* (New York: Bantam, 1979), 2.

36. Greene, *Fabric*, 231.

37. Ibid., 231.

38. Ibid., 228.

39. Ibid., 248.

40. Ibid., 250.

41. Ibid., 15.

42. Greene, *Elegant Universe*, 351.

43. Greene, *Fabric*, 261.

44. Ibid., 265.

45. Ibid., 302.

46. Greene, *Hidden Reality*, 11.

47. Ibid., 66.

48. Ibid., 131.

49. Ibid., 407.

50. Ibid., 410–12, 531.

51. Ibid., 412.

Chapter 6

1. Stephen Hawking, *My Brief History* (New York: Bantam, 2013), 122.

2. Lucy Hawking and Stephen Hawking, *George's Secret Key to the Universe* (New York: Simon & Schuster, 2007), 273.

3. Ibid., 285.

4. Ibid., 98.

5. Ibid., 237.

6. Stephen Hawking, *Black Holes and Baby Universes, and Other Essays* (New York: Bantam, 1994), 29.

7. Stephen Hawking, *A Brief History of Time: From the Big Bang to Black Holes* (New York: Bantam, 1988), 9.

8. Ibid., 92.

9. John Mitchell, "On the Means of Discovering the Distance, Magnitude, &c. of the Fixed Stars, in Consequence of the Diminution of the Velocity of Their Light, in Case Such a Diminution Should Be Found to Take Place in Any of Them, and Such Other Data Should Be Procured from Observations, as Would Be Farther Necessary for That Purpose," *Philosophical Transactions of the Royal Society of London* 74 (1784): 35–57.

10. Hawking, *Brief History*, 85.

11. Ibid., 93–94.

12. Ibid., 85.

13. Ibid., 88.

14. Ibid., 105.

15. Ibid. 112.

16. Ibid., 138.

17. Stephen Hawking, *The Universe in a Nutshell* (New York: Bantam, 2001), 143.

18. Ibid., 144.

19. Ibid., 146.

20. Ibid., 150.

21. Ibid., 150.

22. Ibid., 153.

23. Donald Davidson, "On the Very Idea of a Conceptual Scheme," in *Inquiries into Truth and Interpretation* (Oxford: Clarendon, 1984), 187.

24. Stephen Hawking and Leonard Mlodinow, *The Grand Design* (New York: Bantam, 2010), 79.

25. Stephen Hawking, ed., *Stephen Hawking's* A Brief History of Time: *A Reader's Companion* (New York: Bantam, 1992), 147.

26. Hawking and Mlodinow, *Grand Design*, 159.

27. Ibid., 160.

28. Ibid., 160.

29. Ibid., 159.

30. Ibid., 136.

31. Hélène Mialet, *Hawking Incorporated: Stephen Hawking and the Anthropology of the Knowing Subject* (Chicago: University of Chicago Press, 2012), 22.

32. Hélène Mialet, "The Extended Body of Stephen Hawking," *Interdisciplinary Science Reviews* 37, no. 4 (2012): 256.

33. Mialet, *Hawking Incorporated*, 19.

34. Ibid., 87–89.

35. Ibid., 69.

36. Ibid., 71.

37. Ibid., 24.

38. Jane Hawking, *Music to Move the Stars: A Life with Stephen* (London: Macmillan, 1999), 523–24.

39. Mialet, *Hawking Incorporated*, 25.

40. Robert C. Tucker, "The Theory of Charismatic Leadership," *Daedalus* 97, no. 3 (1968): 731.

41. Mialet, *Hawking Incorporated*, 182.

42. Ibid., 87.

43. Ibid., 44.

44. Ibid., 103.

45. Tucker, "Theory of Charismatic Leadership," 747.

46. Ibid., 742.

47. Karen Walton, "Stephen Hawking: A Cross-Generational Hero," *Physics Teacher* 31 (1993): 148.

48. Walton, "Stephen Hawking," 152.

49. Mialet, *Hawking Incorporated*, 89.

50. Ibid., 228.

51. Patricia Fara, "Newton Lived Here: Sites of Memory and Scientific Heritage," *British Journal for the History of Science* 33, no. 4 (2000): 411.

52. Hawking and Mlodinow, *Grand Design*, 27.

53. "Isaac Newton," Trinity College Chapel website, http://trinitycollegechapel. com//about/memorials/statues/newton.

54. William Wordsworth, *Prelude*, Book 3, ll. 59–63, available online at http://www. bartleby.com/145/ww289.html.

55. Mialet, *Hawking Incorporated*, 148.

56. Hawking, *Brief History*, 193.

Chapter 7

1. http://wveerww.fws.gov/rachelcarson/ (accessed June 24, 2015).

2. Coastal Maine Botanical Gardens, "'Silent Spring' 50th Anniversary," *Wiscasset Newspaper*, August 22, 2012, http://www.wiscassetnewspaper.com/article/silent-spring-50th-anniversary/2055.

3. Quoted in William Souder, *On a Farther Shore: The Life and Legacy of Rachel Carson* (New York: Crown, 2012), 217.

4. Martha Freeman, ed., *Always, Rachel: The Letters of Rachel Carson and Dorothy Freeman, 1952–1964* (Boston: Beacon, 1995), 20.

5. Quoted in Souder, *On a Farther Shore*, 368.

6. Ibid., 335.

7. Ibid.,370.

8. Ibid.,214.

9. Linda J. Lear, *Rachel Carson: Witness for Nature* (New York: H. Holt, 1997), 577.

10. Freeman, *Always* 130.

11. Thomas Nagel, *Mortal Questions* (Cambridge: Cambridge University Press, 1979), 169.

12. Henry Williamson, *Tarka the Otter: His Joyful Waterlife and Death in the Country of the Two Rivers* (New York: G. P. Putnam, 1927), 7.

13. Henry Williamson, *Salar the Salmon* (London: Faber & Faber), 155.

14. Rachel Carson, *Under the Sea Wind* (London: Penguin, 1996), 162.

15. "Background," Henry Williamson Society website, http://www.henrywilliamson. co.uk/bibliography/a-lifes-work/tarka-the-otter#background.

16. Souder, *On a Farther Shore*, 290.

17. Williamson, *Salar*, 37.

18. Carson, *Sea Wind*, 202.

19. Ibid., 200.

20. Kim Heacox, *Antarctica: The Last Continent* (Washington, DC: National Geographic, 1998), 61–64.

21. Rachel Carson, *The Sea Around Us* (New York: New American Library, 1961), 92; see Devra Kunin, trans., *The Passion and Miracles of the Blessed Óláfr*, ed. Carl Phelpstead (Exeter, UK: Short Run, 2008), available online at http://www.vsnrweb-publications.org.uk/Text%20Series/Historia%26Passio.pdf.

22. Carson, *Sea Around Us*, 30.

23. Ibid., 41.

24. Ibid., 118.

25. Ibid., 110.

26. Ibid., 11, my italics.

27. John Lyly, *Euphues: The Anatomy of Wit* and *Euphues and His England*, ed. Morris William Croll and Harry Clemons (London: George Routledge & Sons, 1916), 11.

28. Lear, *Rachel Carson*, 228.

29. Quoted in Ibid., 242.

30. Rachel Carson, *The Edge of the Sea* (New York: New American Library, 1955), 194.

31. Ibid., 195.

32. Ibid., 206–7.

33. Ibid., 238.

34. Ibid., 201.

35. Ibid., 229.

36. Ibid., 226, 220, 206.

37. Ibid., 233.

38. Ibid., 245.

39. Ibid., 218–19.

40. Edwin Diamond, "The Myth of the 'Pesticide Menace,'" *Saturday Evening Post*, September 28, 1963, 16; LeMonte C. Cole, "Rachel Carson's Indictment of the Wide Use of Pesticides," *Scientific American* 207 (December 1962): 176.

41. Rachel Carson, *Silent Spring* (Boston: Houghton Mifflin, 2002), 2.

42. Ibid., 277.

43. Christine Oravec, "An Inventional Archaeology of 'A Fable of Tomorrow,'" in *And No Birds Sing: Rhetorical Analyses of Rachel Carson's* Silent Spring, ed. Craig Waddell (Carbondale: Southern Illinois Press, 2000), 46.

44. Carson, *Silent Spring* 297.

45. Randy Harris, "Other-Words in *Silent Spring*," in Waddell, *And No Birds Sing*, 153.

46. Carson, *Silent Spring*, 109.

47. Ibid.,27–28.

48. Ibid., 25.

49. Ibid., 189.

50. Ibid., 67–68.

51. Ibid., 272–74.

52. Ibid.,17.

53. Ibid., 69.

54. Ibid., 172.

55. Ibid., 80.

56. Ibid., 259.

57. Ibid., 35–36. See "Arsenites in Agriculture," *The Lancet* 275, no. 7116. (January 16, 1960): 178.

58. Carson, *Silent Spring* 193. See V. B. Wigglesworth, "A Case of D.D.T. Poisoning in Man," *British Medical Journal* 1, no. 4397 (April 14, 1945): 517.

59. Kenny Walker and Lynda Walsh, "'No one yet knows what the consequences will be': How Rachel Carson Transformed Scientific Uncertainty into a Site for Public Participation in *Silent Spring*," *Journal of Business and Technical Communication* 26, no. 1 (2012): 23.

60. Carson, *Silent Spring*, 23.

61. Walker and Walsh, "No one yet knows," 27.

62. Carson, *Silent Spring*, 12.

63. Patrick T. O'Shaughnessy, "Parachuting Cats and Crushed Eggs: The Controversy Over the Use of DDT to Control Malaria," *American Journal of Public Health*, November 2008: 1940–48; Henk van den Berg, "Global Status of DDT and Its Alternatives for Use in Vector Control to Prevent Disease," *Environmental Health Perspcetives* 117, no. 11. (November 2009): 1656–63; Sean M Griffing, Dionicia Gamboa, and Venkatachalam Udhayakumar, "The History of 20th Century Malaria Control in Peru," *Malaria Journal* 12 (2013); S. S. Sahu, K. Gunasekaran, H. K. Raju, P. Vanamail, M. M. Pradhan, and P. Jambulingam, "Response of Malaria Vectors to Conventional Insecticides in the Southern Districts of Odisha State," *Indian Journal of Medical Research* 139 (February 2014): 294–300.

64. David Kinkela, *DDT and the American Century: Global Health, Environmental Politics, and the Pesticide that Changed the World* (Chapel Hill: University of North Carolina Press, 2011), 182, 129.

Chapter 8

1. Stephen Jay Gould, *Questioning the Millennium* (New York: Harmony, 1997), 204–5.

2. Philip Morrison, "Banality of IQ," *Scientific American*, March 1982, 30–34.

3. J. Philippe Rushton, "Review," *Society*, March-April 1997, 82.

4. J. Philippe Rushton, "Special Review," *Personality and Individual Differences* 23, no. 1 (1997): 177.

5. Howard Gardner, *Frames of Mind: The Theory of Multiple Intelligences* (New York: Basic Books, 2011).

6. Howard Gardner and S. Moran. "The Science in Multiple Intelligences: A Response to Lynn Waterhouse," *Educational Psychologist* 41 (2006): 230.

7. Charles Spearman, "General Intelligence, Objectively Determined and Measured," *American Journal of Psychology* 15 (1904): 285.

8. Stephen Jay Gould, *The Mismeasure of Man* (New York: Norton 1996), 332.

9. Ibid., 269.

10. Ibid., 341.

11. Robert Cudek and Robert C. MacCallum, *Factor Analysis at 100: Historical Developments and Future Directions* (Mahwah, NJ: Lawrence Erlbaum, 2007), 241.

12. Steve Blinkhorn, "What Skulduggery?," *Nature* 296 (1982): 506.

13. Peter Novick, *The Holocaust in American Life* (Boston: Houghton-Mifflin, 1999), 44.

14. Alexandra Molnar, "History of Italian Immigration," 2010, https://www.mtholyoke.edu/~molna22a/classweb/politics/Italianhistory.html.

15. Mark Snyderman and R. J. Herrnstein, "Intelligence Tests and the Immigration Act of 1924," *American Psychologist* (September 1983): 986–95.

16. Gould, *Mismeasure*, 323.

17. Franz Samelson, "Intelligence and Some of Its Testers," *Science* 215 (1982): 656–67; see also John Lawson and Harold Silver, *A Social History of Education in England* (London: Methuen, 1973), 393–97.

18. Jason E. Lewis, David DeGusta, Marc R. Meyer, Janet M. Monge, Alan E. Mann, and Ralph L. Holloway, "The Mismeasure of Science: Stephen Jay Gould versus Samuel George Morton on Skulls and Bias," *PloS* Biology, June 7, 2011 http://dx.doi.org/10.1371/journal.pbio.1001071; John S. Michael, "A New Look at Morton's Craniological Research," *Current Anthropology* 29, no. 2 (1988): 349–54; Michael Weisberg, "Remeasuring Man," *Evolution and Development* 16, no. 3 (2004): 166–78.

19. Stephen Jay Gould, "Morton's Ranking of Races by Cranial Capacity" *Science*, n.s., 200, no. 4341 (1978): 508.

20. Weisberg, "Remeasuring Man," 177.

21. Leigh Van Valen, "Brain Size and Intelligence in Man," *American Journal of Physical Anthropology* 40 (1974): 419.

22. Rushton, "Review," 81.

23. Van Valen, "Brain Size," 420.

24. Gould, *Mismeasure*, 27.

25. Ibid., 36.

26. Stephen Jay Gould, *Time's Arrow, Time's Cycle: Myth and Metaphor in the Discovery of Geological Time* (Cambridge, MA: Harvard University Press, 1987), 2.

27. Thomas Burnet, *Sacred Truth*, vol. 1 (Glasgow: R. Urie, 1753), 81–82.

28. Gould, *Time's Arrow, Time's Cycle*, 41.

29. H. W. Turnbull, ed., *The Correspondence of Isaac Newton*, vol. 7 (Cambridge: Cambridge University Press, 1959), 334

30. Burnet, *Sacred Truth*, 38.

31. Gould, *Time's Arrow, Time's Cycle*, 63.

32. Stuart A. Baldwin, "Charles Lyell," *Geology Today*, May-June 1998, 113–15.

33. Charles Lyell, *Principles of Geology* (New York: D. Appleton, 1854), 53.

34. Ibid., 123.

35. Ibid., 113.

36. Ibid., 152.

37. Stephen Jay Gould, *Wonderful Life: The Burgess Shale and the Nature of History* (New York: W. W. Norton, 1989), 14.

38. Ibid., 65.

39. Ibid., 52.

40. Ibid., 83.

41. Ibid., 241.

42. Ibid., 14.

43. Ibid., 283.

44. Ibid., 25.

45. Ibid., 212.

46. Ibid., 288.

47. H. B. Whittington, "The Lobopod Animal *Aysheaia pedunculata* Walcott, Middle Cambrian Burgess Shale, British Columbia," *Philosophical Transactions of the Royal Society London B* 284 (1984): 195.

48. Quoted in Gould, *Wonderful Life*, 118.

49. Ibid., 100.

50. Gardner, *Frames of Mind*, 185.

51. Robert H. McKim, *Experiences in Visual Thinking* (Monterey, CA: Brooks/Cole, 1972), 8.

52. H. B. Whittington, "The Lobopod animal *Aysheaia pedunculata* Walcott, Middle Cambrian, Burgess Shale, British Columbia," *Philosophical Transactions of the Royal Society of London B*, 284, no. 1000 (1978): 171.

53. Gould, *Wonderful Life*, 116.

54. Ibid., 46

55. Ibid., 47.

56. Ibid., 228.

57. Simon Conway Morris, "Showdown on the Burgess Shale: The Challenge," *Natural History* 107, no. 10 (1998): 48–55.

58. Derek E. G. Briggs, "Extraordinary Fossils Reveal the Nature of Cambrian Life: A Commentary on Whittington (1975) 'The Enigmatic Animal *Opabinia regalis*, Middle Cambrian, Burgess Shale, British Columbia,' " *Philosophical Transactions of the Royal Society B* 370 (April 19, 2015): 20140313.

59. Niles Eldredge and S. J. Gould, "Punctuated Equilibria: An alternative to Phyletic Gradualism," in *Models in Paleobiology*, ed. T. J. M. Schopf (San Francisco: Freeman Cooper, 1972), 82–115.

60. S. J. Gould and R. C. Lewontin, "The Spandrels of San Marco and the Panglossian Paradigm: A Critique of the Adaptationist Programme," *Proceedings of the Royal Society of London B* 205, no. 1161 (1979): 581–98.

Chapter 9

1. Stephen Jay Gould, *The Flamingo's Smile* (New York: W. W. Norton, 1985), 167.

2. Stephen Jay Gould, *Bully for Brontosaurus* (New York: W. W. Norton, 1991), 12.

3. Adam Smith, *Essays on Philosophical Subjects*, ed. W. P. D. Wightman and J. G. Bryce (Indianapolis: Liberty Fund, 1982), 40.

4. Gould, *Bully*, 60.

5. Ibid., 66.

6. Ibid., 78.

7. Stephen Jay Gould, *Dinosaur in a Haystack* (New York: Harmony, 1995), 388.

8. Ibid., 392.

9. Ibid., 393.

10. Ibid., 396-97.

11. Ibid., 149.

12. Ibid., 151.

13. Peter Schulte, Laia Alegret, Ignacio Arellas, et al. "The Chicxulub Asteroid Impact and Mass Extinction at the Cretaceous-Paleogene Boundary." *Science* 327 (2010): 1214–18.

14. Gould, *Dinosaur* ,157.

15. Ernst Mayr, "Speciational Evolution or Punctuated Equilibria," in *The Dynamics of Evolution*, ed. Albert Somit and Steven Peterson (Ithaca, NY: Cornell University Press, 1992), 48.

16. Robert Wright, "Intelligence Test," review of Stephen Jay Gould's *Wonderful Life, New Republic*, January 29, 1990, 28.

17. Gould, *Dinosaur*, 140.

18. Ibid., 142.

19. Ibid., 143.

20. Ibid., 103.

21. Stephen Jay Gould, *Eight Little Piggies: Reflections in Natural History* (New York: W. W. Norton, 1993), 279.

22. Ibid., 280.

23. Ibid., 280.

24. Ibid., 281.

25. Ibid., 283.

26. Gould, *Dinosaur*, 126.

27. Ibid., 128.

28. R. G. Collingwood, *The Idea of History* (New York: Oxford University Press, 1956), 237.

29. Stephen Jay Gould, *Leonardo's Mountain of Clams and the Diet of Worms: Essays on Natural History* (New York: Three Rivers, 1998), 31.

30. Stephen Jay Gould, *I Have Landed: The End of a Beginning in Natural History* (New York: Harmony, 2002), 293.

31. Ibid., 295.

32. Ibid., 296.

33. Ibid., 303.

34. Gould, *Leonardo's Mountain*, 325.

35. Ibid., 326.

36. Ibid., 328.

37. Quoted in Ibid., 329.

38. Quoted in Ibid., 332.

Chapter 10

1. Ed Douglas, "Steven Pinker: The Mind Reader," *Guardian*, November 5, 1999.

2. Ibid.

3. Ibid.

4. Christopher Longuet-Higgins, "The Talking Ape," *Nature* 368 (March 24, 1994): 360–61.

5. Steven Pinker, interview by Dan Schneider, *Cosmoetica*, July 13, 2007, http://www.cosmoetica.com/DSI4.htm.

6. Ibid.

7. Benjamin Thorpe, ed., *The Homilies of the Anglo-Saxon Church*, vol. 1 (London: Richard and John E. Taylor, 1844), 185, available online at http://www.gutenberg.org/files/38334/38334-h/38334-h.htm.

8. Steven Pinker, *The Language Instinct* (New York: Harper Perennial, 1995), 183.

9. Ibid., 162.

10. Ibid., 171.

11. Ibid., 172.

12. Ibid., 172.

13. Ibid., 191.

14. Ibid., 398.

15. Noam Chomsky, review of *Verbal Behavior* by B. F. Skinner, *Language* 35, no. 1 (1959): 55–56.

16. J. L. McClelland, D. E. Rumelhart, and the PDP Research Group, "On Learning the Past Tense of English Verbs," in *Parallel Distributed Processing: Explorations in the Microstructure of Cognition*, vol. 2 (Cambridge, MA: MIT Press, 1986), 267.

17. Steven Pinker, *Words and Rules: The Ingredients of Language* (New York: Harper Perennial, 2000), 110.

18. Ibid., 117.

19. James M. McClelland and Karalyn Patterson, "Rules and Connections in Past-Tense Inflections: What Does the Evidence Rule Out?" *Trends in Cognitive Science* 6, no. 11 (November 2002): 471.

20. Franklin Chang, "Learning to Order Words: A Connectionist Model of Heavy NP Shift and Accessibility Effects in Japanese and English," *Journal of Memory and Language* 61 (2009): 392.

21. Steven Pinker, *The Stuff of Thought: Language as a Window into Human Nature* (New York: Viking, 2007), 95.

22. Stephen Schiffer, Review of *Concepts: Where Cognitive Science Went Wrong* by Jerry A. Fodor, *Times Literary Supplement*, June 26, 1998, 15.

23. Jerry A. Fodor, *Concepts: Where Cognitive Science Went Wrong* (Oxford: Clarendon, 2009).

24. Edmund Gettier, "Is Justified True Belief Knowledge?" *Analysis* 23, no. 6. (June 1963): 121–23.

25. Pinker, *Stuff of Thought*, 89–151.

26. Steven Pinker, *How the Mind Works* (New York: W. W. Norton, 2009), 13.

27. Ibid., 281.

28. Ibid., 282.

29. Ibid., 233.

30. Ibid., 561 (his emphasis)

31. Ibid., 50.

32. Thomas Nagel, *Mind and Cosmos: Why the Materialist Neo-Darwinian Conception of Nature Is Almost Certainly False* (New York: Oxford, 2012), 68.

33. Pinker, *How the Mind Works*, 4.

34. Jerry A. Fodor, *The Mind Doesn't Work That Way: The Scope and Limits of Computational Psychology* (Cambridge, MA: MIT Press, 2000), 37.

35. Indeed, recent findings indicate that men are innately better at chess, a finding that differences in participation rates cannot explain; see Robert Howard, "Gender Differences in Intellectual Performance Persist at the Limits of Individual Capabilities," *Journal of Biosocial Science* 46, no. 3 (2013): 386–404; Michael Knapp, "Are Participation Rates Sufficient to Explain Gender Differences in Chess Performance?" *Proceedings of the Royal Society of London B* 277 (2010): 2269–70.

36. Lawrence H. Summers, "Remarks at NBER Conference on Diversifying the Science & Engineering Workforce," January 14, 2005, available online at http://www.harvard.edu/president/speeches/summers_2005/nber.php.

37. *The Happy Feminist* (blog), February 22, 2006, http://happyfeminist.typepad.com/happyfeminist/2006/02/larry_summers_r.html.

38. Steven Pinker, *The Blank Slate: The Modern Denial of Human Nature* (New York: Viking, 2002).

39. Ibid., 346.

40. Ibid., 355.

41. Ibid., 367

42. Ibid., 370.

43. Brenda J. Benson, Carol L. Gohm, and Alan M. Gross, "College Women and Sexual Assault: The Role of Sex-Related Alcohol Expectancies," *Journal of Family Violence* 22 (2007): 341–51.

44. Steven Pinker, *The Better Angels of Our Nature: Why Violence Has Declined* (New York: Penguin, 2011), 406.

45. Pinker, *How the Mind Works*, 537.

46. Randall Jarrell, *Poetry and the Age* (New York: Vintage, 1955), 10.

47. Pinker, *Blank Slate*, 415.

48. Iris Berent, Steven Pinker, Joseph Tzelgo, Uri Bibi, and Liat Goldfarb, "Computation of Semantic Number from Morphological Information," *Journal of Memory and Language* 53 (2005): 354.

49. Pinker, *Better Angels*.

50. Steven Pinker, "Frequently Asked Questions about *The Better Angels of Our Nature: Why Violence Has Declined*," Steven Pinker website, http://stevenpinker.com/pages/frequently-asked-questions-about-better-angels-our-nature-why-violence-has-declined.

51. I have rearranged the order of Pinker's original table from a given cause's unadjusted to adjusted rank.

52. "List of Wars and Anthropogenic Disasters by Death Toll," *Wikipedia*, http://en.wikipedia.org/wiki/List_of_wars_and_anthropogenic_disasters_by_death_toll.

53. Pinker, *Better Angels*, 668.

54. James R. Flynn, "Searching for Justice: The Discovery of IQ Gains over Time," *American Psychologist* 54, no. 1 (January 1999): 7.

Chapter 11

1. Richard Dawkins, *An Appetite for Wonder: The Making of a Scientist* (New York: HarperCollins, 2013), 38.

2. Ibid., 142.

3. Longinus, *On the Sublime*, trans. W. H. Fyfe (Cambridge, MA: Harvard University Press, 1995), 181.

4. Richard Dawkins, *The Selfish Gene* (Oxford: Oxford University Press, 1998), 12.

5. Adam Smith, *The Glasgow Edition of the Works and Correspondence of Adam Smith*, vol. 3, ed. W. P. D. Wightman and J. C. Bryce (Oxford: Clarendon, 1983), 75.

6. Ibid., 33.

7. Adam Smith, *Theory of Moral Sentiments* (London: George Bell & Sons, 1875), 113.

8. Charles Darwin, *The Origin of Species* (Cambridge, MA: Harvard University Press, 1964), 74–75.

9. G. H. Hardy, "Mendelian Proportions in a Mixed Population," *Science*, New Series, 28 (July 10, 1908): 49.

10. G. H. Hardy, *A Mathematician's Apology* (Cambridge: Cambridge University Press, 1967), 129.

11. Richard Dawkins, *The Selfish Gene* (Oxford: Oxford University Press, 1998), 69.

12. Ibid., 69.

13. Ibid., 282.

14. Ibid., 282.

15. John Maynard Smith and G. A. Parker, "The Logic of Asymmetric Contests," *Animal Behavior* 24 (1976): 159–75.

16. Dawkins, *Selfish Gene*, 76.

17. Ibid., 76.

18. Ibid., 77.

19. John Maynard Smith, "Evolution and the Theory of Games: In Situations Characterized by Conflict of Interest, the Best Strategy to Adopt Depends on What Others Are Doing," *American Scientist* 64 (January–February 1976): 43–44.

20. H. Jane Brockman, Alan Grafen, and Richard Dawkins, "Evolutionary Stable Nesting Strategy in a Digger Wasp," *Journal of Theoretical Biology* 77 (1979): 476.

21. Richard Dawkins, *The Extended Phenotype: The Gene as a Unit of Selection*, rev. ed. (New York: Oxford University Press, 1999), 131.

22. Ibid., 129–30.

23. John Maynard Smith, "Genes and Memes," *London Review of Books*, 4, no. 2 (February 4, 1982): 3–4.

24. Mary Midgley, "Selfish Genes and Social Darwinism," *Philosophy*, 58 (1983): 372.

25. Dawkins, *Selfish Gene*, xiii.

26. David Hull, *Science as a Process: An Evolutionary Account of the Social and Conceptual Development of Science* (Chicago: University of Chicago Press, 1988), 413.

27. David Hull, review of *The Selfish Gene* by Richard Dawkins, *Nature* 342 (1989): 319.

28. Richard Dawkins, *The Selfish Gene*, rev. ed. (Oxford: Oxford University Press, 2006), 38.

29. Ibid., 19–20.

30. Ibid., 270–71.

31. Ibid., 277.

32. Ibid., 277.

33. Ibid., 59.

34. Michael Pearson, "Marine Who Took Grenade Blast for Comrade Receives Medal of Honor," *CNN*, June 19, 2014, http://www.cnn.com/2014/06/19/us/medal-of-honor-carpenter/index.html.

35. Dawkins, *Selfish Gene*, 33.

36. Ibid., 62.

37. Richard Dawkins, *Climbing Mount Improbable* (New York: Norton, 1996), 255.

38. Peter K. Dearden and Michael Akam, "Early Embryo Patterning in the Grasshopper, *Schistocerca gregaria*: Wingless, Decapentaplegic and Caudal Expression." *Development* 128 (2001): 3435–44.

39. Marc Kirschner and John Gerhart, "Evolvability," *Proceedings of the National Academy of Science, USA* 95 (1998): 8424; see also Gunter P. Wagner and Lee Altenberg, "Complex Adaptations and the Evolution of Evolvability," *Evolution* 50 (1996): 967–76.

40. Nathan M. Young, Günter P. Wagner, and Benedikt Hallgrímsson, "Development and the Evolvability of Human Limbs," *Proceedings of the National Academy of Science, USA* 207 (2010): 3400–3405.

41. Alex Mesoudi, Andrew Whiten, and Kevin L. Laland, *Behavioral and Brain Sciences* 29 (2006): 329–83.

42. Dawkins, *Selfish Gene*, 199–200.

43. Mary Midgley, "Gene Juggling," *Philosophy* 54 (1979): 457.

44. Dawkins, *Selfish Gene*, 192.

45. Alan G. Gross, Joseph E. Harmon, and Michael Reidy, *Communicating Science* (New York: Oxford University Press, 2006).

46. Myron C. Baker and David E. Gammon, "Vocal Memes in Natural Populations of Chickadees: Why Do Some Memes Persist and Others Go Extinct?" *Animal Behaviour* 75 (208): 279–89.

47. Gonçalo C. Cardoso and Jonathan W. Atwell, "Directional Cultural Change by Modification and Replacement of Memes," *Evolution* 65 (2011): 295–300.

48. Adrian L. O'Loghlen, Vincenzo Ellis, Devin R. Zaratzian, Loren Merrill, and Stephen I. Rothstein, "Cultural Evolution and Long-Term Song Stability in a Dialect Population of Brown-Headed Cowbirds," *Condor* 113 (2011): 449–61.

49. Dominique A. Potvin and Kirsten M. Parris, "Song Convergence in Multiple Urban Populations of Silvereyes (Zosterops lateralis)," *Ecology and Evolution* 2 (2012): 1977–84.

50. Cardoso and Atwell, "Directional Cultural Change," 299.

51. Dawkins, *Appetite for Wonder*, 280.

52. Richard Dawkins, *The Blind Watchmaker: Why the Evidence of Evolution Reveals a Universe without Design* (New York: Norton, 1996), 56–57.

53. Ibid., 72.

Chapter 12

1. Charles Darwin, Letter 176, available online at http://www.darwinproject.ac.uk/entry-176.

2. Edward O. Wilson, *Biophilia: The Human Bond with Other Species* (Cambridge, MA: Harvard University Press, 1984), 27

3. Edward O. Wilson, *Naturalist* (New York: Warner Books, 1995), 306.

4. Ibid., 306.

5. Ibid., 304.

6. Richard Lewontin, "Sleight of Hand," *The Sciences* (July/August 1981): 26.

7. Wilson, *Naturalist*, 346.

8. Ibid., 349; Ullica Segerstrale, *Defenders of the Truth: The Battle for Science in the Sociological Debate and Beyond* (New York: Oxford University Press, 2000), 19–24.

9. Wilson, *Naturalist*, 342,

10. Ibid., 299–306, 341–47.

11. Wilson, *Biophilia*, 17.

12. Edward O. Wilson, "Karl von Frisch and the Magic Well," *Science* 159, no. 3817 (1968): 865.

13. Bert Hölldobler and Edward O. Wilson, *The Leafcutter Ants: Civilization by Instinct* (New York: W. W. Norton, 2010), 115.

14. Bert Hölldobler and Edward O. Wilson, *Journey to the Ants: A Story of Scientific Exploration* (Cambridge, MA: Harvard University Press, 1994), 48.

15. Hölldobler and Wilson, *Leafcutter Ants*, 125.

16. Hölldobler and Wilson, *Journey*, 115.

17. Hölldobler and Wilson, *Leafcutter Ants*, 63.

18. Edward O. Wilson, *Anthill: A Novel* (New York: W. W. Norton, 2011), 201.

19. Hölldobler and Wilson, *Journey*, 61–62.

20. Ibid., 58.

21. Elizabeth Allen et al. "Against 'Sociobiology,'" *New York Review of Books*, November 13, 1975.

22. Edward O. Wilson, *Sociobiology: The New Synthesis* (Cambridge, MA: Harvard University Press, 1975), 564.

23. Edward O. Wilson, *Consilience: The Unity of Knowledge* (New York: Vintage, 1999), 233.

24. Roger Scruton, *Understanding Music: Philosophy and Interpretation* (London: Continuum, 2009), 11.

25. Scruton, *Understanding Music*, 38.

26. William James, *The Will to Believe and Other Essays in Popular Philosophy* (New York: Longman Green, 1903), 76.

27. William Youngren, fanfarearchive.com/articles/atop/04_6/0460300.az_BEETH-OVEN_Piano_Sonatas_6.html. Thanks to Randy Allen Harris.

28. Daniel Barenboim, "Beethoven and the Quality of Courage," *New York Review of Books*, April 3, 2013.

29. Peter Burwasser, fanfarearchive.com/articles/atop/31_2/3120870.az_BEETH-OVEN_Piano_Sonatas_16.htm.

30. Janet Schmalfeldt, "One More Time on Beethoven's 'Tempest,' from Analytic and Performance Perspectives: A Response to William E. Caplin and James Hepokoski," *Music Theory Online* 16, no. 2 (2010), http://www.mtosmt.org/issues/mto.10.16.2/mto.10.16.2.schmalfeldt3.html.

31. Schmalfeldt, "One More Time."

32. Edward O. Wilson, *On Human Nature* (Cambridge, MA: Harvard University Press, 1978), 167.

33. Bernard Williams, *Ethics and the Limits of Philosophy* (Cambridge, MA: Harvard University Press, 1985), 111.

34. Bernard Williams, *Problems of Self* (Cambridge: Cambridge University Press, 1973), 175.

35. Bernard Williams, *Moral Luck* (Cambridge: Cambridge University Press, 1981), 28.

36. Williams, *Moral Luck*, 38.

37. Williams, *Ethics*, 44; see also Marshall D. Sahlins, "The Use and Abuse of Biology," in *The Sociobiology Debate: Readings on the Ethical and Scientific Issues Concerning Sociobiology*, ed. Arthur L. Caplan (New York: Harper & Row, 1978), 424–27.

38. Maciej Chudek and Joseph Henrich, "Culture-Gene Coevolution, Norm-Psychology, and the Emergence of Human Prosociality," *Trends in Cognitive Sciences* 15, no. 5 (2011): 223.

39. Peter DeScioli and Robert Kurzban, "Mysteries of Morality," *Cognition* 112 (2009): 281–99.

40. Edward O. Wilson, *The Future of Life* (New York: Vintage, 2002), 5.

41. Ibid., 20.

42. Edward O. Wilson, *The Diversity of Life* (Cambridge, MA: Harvard University Press, 1992), 35.

43. Wilson, *Diversity*, 20.

44. Ibid., 223.

45. Ibid., 224.

46. Wilson, *Future*, 53.

47. Wilson, *Diversity*, 164.

48. Ibid., 202.

49. William F. Laurance et al. "The Fate of Amazonian Forest Fragments: A 32-Year Investigation," *Biological Conservation* 144 (2011): 65.

50. Stephen R. Kellert and Edward O. Wilson, eds., *The Biophilia Hypothesis* (Washington, DC: Island, 1993), 263.

51. Ibid., 263.

52. Ibid., 60.

53. Ibid., 63.

54. W. M. Mann, "Stalking Ants, Savage and Civilized," *National Geographic* 66, no 2 (1934): 177.

55. Wilson, *Naturalist*, 60.

56. Wilson, *Biophilia*, 16.

Chapter 13

1. Leo Tolstoy, *Anna Karenina*, trans. Joel Carmichael (New York: Bantam, 2006), 598.

2. C. P. Snow, foreword to G. H. Hardy, *A Mathematician's Apology* (Cambridge: Cambridge University Press, 1967), 30–37.

3. Sharon Traweek, *Beamtimes and Lifetimes: The World of High Energy Physics* (Cambridge, MA: Harvard University Press, 1988), 79.

4. Steven Weinberg, "Change Course," *New York Times*, September 5, 2009, http://www.nytimes.com/2009/09/06/opinion/06weinberg.html?_r=0.

5. Joseph C. Hermanowicz, *Lives in Science: How Institutions Affect Academic Careers* (Chicago: University of Chicago Press, 2009), 17.

6. Steven Weinberg, *Dreams of a Final Theory: The Search for the Fundamental Laws of Nature* (New York: Pantheon, 1992), 245.

7. Edward O. Wilson, *The Meaning of Human Existence* (New York: Liveright, 2014.), 149.

8. Anselm, *Proslogium*, chapters 2 and 3, available online at http://www.fordham.edu/halsall/basis/anselm-proslogium.asp.

9. Richard Dawkins, *The God Delusion* (Boston: Houghton Mifflin, 2006), 104.

10. P. J. McGrath, "The Refutation of the Ontological Argument," *Philosophical Quarterly* 40 (1990): 195–212; Peter Millican," The One Fatal Flaw in Anselm's Argument," *Mind*, n.s., 113 (2004): 437–76; J. William Forgie, "How Is the Question 'Is Existence a Predicate?' Relevant to the Ontological Argument?" *International Journal*

of the *Philosophy of Religion* 64 (2006): 117–33; William L. Rowe, "Alvin Plantinga on the Ontological Argument," *International Journal of the Philosophy of Religion* 65 (2009): 87–92.

11. Dawkins, *God Delusion*, 104.

12. Ibid., 251.

13. Ibid., 254.

14. Ibid., 51.

15. Ibid., 274–78.

16. Ibid., 274–75.

17. Rosemary Woolf, "The Effect of Typology on the English Mediaeval Plays of Abraham and Isaac," *Speculum* 32, no. 4 (October 1957): 805–25.

18. Augustine, *Reply to Faustus the Manichaean* 22.73, available online at http://www.gnosis.org/library/contf1.htm.

19. Augustine, *Reply to Faustus* 12.25.

20. *The Chester Cycle Play IV (4): Abraham and Isaac*, available online at http://machias.edu/faculty/necastro/drama/chester/play_04.html.

21. Quoted in Jürgen Habermas, *Religion and Rationality: Essays in Reason, God, and Modernity*, ed. Eduardo Mendieta (Cambridge, MA: MIT Press, 2002), 105.

22. Erich Auerbach, *Mimesis: The Representation of Reality in Western Literature*, trans. Willard R. Trask (Princeton, NJ: Princeton University Press, 2003), 14–15.

23. Jack Miles, *God: A Biography* (New York: Vintage, 1995), 58–61.

24. Søren Kierkegaard, *Fear and Trembling; and The Sickness unto Death*, trans. Walter Lowrie (New York: Doubleday, 1954), 30.

25. Kierkegaard, *Fear and Trembling*, 41.

26. Ibid., 128.

27. *Chester Cycle Play IV*.

28. Robert Craft, liner notes to *Stravinsky the Composer*, vol. 6, Musicmasters, CD, 1995, 10

29. Dawkins, *God Delusion*, 283.

30. Richard Dawkins, *The Selfish Gene* (Oxford: Oxford University Press, 2016), 139.

31. A. I Sabra. "Situating Arabic Science: Locality versus Essence," *Isis* 87, no. 4 (December 1996): 669, 661.

32. Habermas, *Religion and Rationality*, 147.

33. See David Lindberg, ed., *Science in the Middle Ages* (Chicago: University of Chicago Press, 1978), and Edward Grant, *Physical Science in the Middle Ages* (Cambridge: Cambridge University Press, 1977).

34. Grant, *Physical Science*, 20–35.

35. Thomas S. Kuhn, *The Copernican Revolution: Planetary Astronomy in the Development of Western Thought* (Cambridge, MA: Harvard University Press, 1957), 192.

36. See Alvin Plantinga, *Where the Conflict Really Lies: Science, Religion, and Naturalism* (Oxford: Oxford University Press, 2011), and Alvin Plantinga and Daniel C. Dennett, *Science and Religion: Are They Compatible?* (New York: Oxford University Press, 2011).

37. Thomas Nagel, *Mind and Cosmos: Why the Materialist Neo-Darwinian Conception of Nature Is Almost Certainly False* (New York: Oxford University Press, 2012), 123.

38. Nagel, *Mind and Cosmos*, 107. See also "Panpsychism" in Thomas Nagel, *Mortal Questions* (Cambridge: Cambridge University Press), 181–95.

39. Edward J. Larson, *Trial and Error: The American Controversy over Creation and Evolution* (New York: Oxford University Press, 1989), 162–66.

40. "In U. S., 46% Hold Creationist Views of Human Origins," Gallup, 2012, http://www.gallup.com/poll/155003/Hold-Creationist-View-Human-Origins.aspx.

41. "New Survey Results on Evolution," Virginia Commonwealth University, May 27, 2010, http://ncse.com/news/2010/05/new-survey-results-evolution-005542.

42. "Evolution, Creationism, Intelligent Design," Gallup, 2013, http://www.gallup.com/poll/21814/Evolution-Creationism-Intelligent-Design.aspx.

43. "Ten Major Court Cases about Evolution and Creationism," National Center for Science Education, February 14, 2001, http://ncse.com/taking-action/ten-major-court-cases-evolution-creationism.

44. Michael B. Berkman, Julianna Sandell Pacheco, and Eric Plutzer, "Evolution and Creationism in Americas Classrooms: A National Portrait" *PLoS Biology* 6, no. 5 (May 2008), e124.

45. Gallup, 2013.

46. RationalWiki, s.v. "Council of Europe 2007 Resolution on the Teaching of Creationism," http://rationalwiki.org/wiki/Council_of_Europe_2007_resolution_on_the_teaching_of_creationism; Council of Europe Parliamentary Assembly, vote on resolution, "The Dangers of Creationism in Education" (Doc. 11375), http://assembly.coe.int/nw/xml/Votes/DB-VotesResults-EN.asp?VoteID=674&DocID=12120&MemberID=5634.

47. Glen Branch, "Teaching Evolution in Turkey," National Center for Science Education, June 25, 2013, http://ncse.com/news/2013/06/teaching-evolution-turkey-0014883.

48. "Religious and Social Attitudes of UK Christians in 2011," http://www.ipsos-mori.com/Assets/Docs/Polls/ipsos-mori-religious-and-social-attitudes-topline-2012.pdf; "Project Darwin Omnibus—Great Britain 2009," http://www.ipsos-mori.com/Assets/Docs/Polls/poll-darwin-survey-shows-international-consensus-on-acceptance-of-evolution.pdf; "Darwin Survey Reveals Divided Britain in Attitudes Toward Evolution," 2009, http://www.britishcouncil.org.

49. "Teachers Dismiss Calls for Creationism to Be Taught in School Science Lessons," 2008, http://www.ipsos-mori.com/researchpublications/researcharchive/2315/Teachers-Dismiss-Calls-For-Creationism-To-Be-Taught-In-School-Science-Lessons.aspx.

50. Ray Jayawardhana, "Listen Up, It's Neutrino Time," *New York Times*, December 23, 2013.

51. Ronald L. Numbers, *The Creationists: The Evolution of Scientific Creationism* (Berkeley: University of California Press, 1993), 247.

52. Edward J. Larson and Larry Witham, "Leading Scientists Still Reject God," *Nature* 394 (July 23, 1998): 313–14.

53. Elaine Howard Eklund, Jerry Z. Park, and Elizabeth Long, "Scientists and Spirituality," *Sociology of Religion* 72, no. 3 (2011): 262.

54. Eklund, Park, and Long, "Scientists and Spirituality," 265.

55. Richard E. Smalley, "Discovering the Fullerenes," Nobel Lecture, December 7, 1996, http://www.nobelprize.org/nobel_prizes/chemistry/laureates/1996/smalley-lecture.pdf, 89.

56. Robert Boyle, *Free Inquiry into the Vulgarly Received Notion of Nature*, (London: H. Clark, 1686), 47, quoted in Steven Shapin and Simon Shaffer, *Leviathan and the Air-Pump: Hobbes, Boyle, and the Experimental Life* (Princeton, NJ: Princeton University Press, 1985), 202.

57. Richard P. Feynman, *What Do You Care What Other People Think? Further Adventures of a Curious Character* (New York: Bantam, 1989), 270.

58. Dawkins, *Selfish Gene*, 278.

INDEX